The New Business of Design

The Forty-fifth
International Design
Conference in Aspen

John Kao, Program Chair

Allworth Press, New York

745.2
I61n

Published by Allworth Press, an imprint of Allworth Communications, Inc.
10 East 23rd Street, New York, NY 10010

Front cover by Milton Glaser

Typography by Sharp Des!gns

ISBN: 1-880559-38-2

Library of Congress Catalog Card Number: 95-76691

Co Contents nts

I. New Business

International Design Conference in Aspen: Welcome
Richard Farson

I'm Richard Farson, president of the IDCA, and on behalf of the Board of Directors I'd like to welcome all of you to our forty-fifth conference.

This conference has always addressed vital issues that confront designers, and this year is surely no exception. Nothing could be more important for designers to examine than the ways in which the *meaning* of design is changing. The design professions are being radically reshaped. New design disciplines are emerging, and the problems and opportunities created by these events are changing the relationship between business and design. To explore these issues in depth, we have planned a more focused and more intensive conference.

We intend to examine the ways in which design is exploding in new directions and accommodating exciting—and to some, frightening—new technologies. And, yes, to look at the ways in which designers are struggling and sometimes suffering over these events. Can designers find new and more rewarding roles in business and in society? And if so, how can they use those positions to make a better world? That, after all, is why this conference exists. We hope this will be a watershed conference for you and for the IDCA as, together, we try to find new meaning in our work.

To guide us through this exploration we have chosen a most remarkable, broad-ranging, accomplished, and distinguished individual. The chair of our program teaches creativity and entrepreneurship at the Harvard Business School. He has started, and now heads, several corporations and consults with many others. He brings a depth of understanding to his knowledge of business that is most appropriate to this task. But he's much more. He's trained as a physician and as a psychiatrist at Harvard and Yale. He's an authority on Chinese medicine, and, especially impressive to me, he's a first-rate classical and jazz pianist. Please join me in welcoming the chair of our 1995 conference, John Kao.

New Business: Redefining the Idea of Design
John Kao

Re-invention, *re*-creation, *re*-definition, *re*-packaging, *re*-purposing. It
seems like *re* has become the new mantra of the age. With apologies to
those who meditate, we may have to retire *om* from our vocabulary and
substitute *re* instead.

Re poses a big challenge for us. Many of us don't quite know whether to
get involved in starting a *re*volution or to simply sit back and *re*lax. I
would go further. The spirit of *re* challenges us in terms of the subject
that we're here to discuss at this particular conference. It suggests a
need for change and a discontinuity with the past that can energize us if
we use it to guide us in a suitable direction over the next few days. It
was what got me to sign on to do this job. My first IDCA experience was
last year, so I'm a relative newcomer.

Having now fourteen years at Harvard Business School, I've seen
firsthand what is going on out in the business world. As I made my
excursion through the world of design, it became clear to me that there
were puzzling discontinuities in terms of the relationship between
design and business. It was almost as if I had encountered two distinct
tribes, with two distinct sets of concepts and two distinct languages for
describing reality—leading to a *Rashomon* kind of phenomenon in
terms of looking at the same thing but describing it in profoundly
different ways.

Frankly, I was puzzled by the tone of polarization that I heard when I
talked to people about the relationship between design and business.

To be specific, I routinely saw people from the business world who
seemed to be dramatically out of touch with what designers actually do
and who were not inclined to provide an appropriate seat at the table for
design to allow discussion of the integration of design and business. By
the same token, I had talked with many designers, some of whom I felt
were out of touch with the core needs of business. So, the relationship
between design and business was like a marriage, but a bit of an uneasy

9

marriage. It was both puzzling and stimulating to my curiosity. It was what brought me, in a sense, to this evening and this conference.

When I was out there looking around, there were also signs of hope. There were interesting corporate environments that had seamlessly and very persuasively effected an integration between design and business that seemed almost second nature. I would walk into some of these companies, and it would be almost what I call the "Twilight Zone" effect. There would be something so striking about this integration that I could just hear that Rod Serling music and think, "Boy, something very interesting is going on here." So there are examples of practice at the leading edge. And that is part of what the effort to design this conference has been about: to bring examples from the leading edge here for our collective consideration.

Let me talk a little bit more about this relationship, about what I saw on each side of the fence. On the business side, to say that business needs to think about design is a no-brainer. Businesses all over the world are searching for answers as never before. To get a flavor of this, you just have to go into a bookstore to see the volume of books offering method- ologies and metaphors for how to lead organizations into the twenty-first century. I would argue that design is fundamental to that passage, but design in a different sense from that to which we were accustomed. I would argue that there are a lot of very interesting, important issues being faced by the business world today that are also design issues. Business needs help.

Look at the question of organizational design. I have never seen so many metaphors flying around to describe an organization as I have over the last couple of years. You see descriptions of an organization as a blue- berry pancake, or a greenhouse or a birch forest or a thousand trees or a pepperoni pizza or a rugby team or even a jazz combo. Whenever you hear that degree of metaphorical language, you know that knowledge is up for grabs. You know that knowledge is up for grabs in terms of designing organizations when you see the proliferation of conferences and study groups and soon-to-be-published books—be warned—on the twenty-first-century organization. The questions become, "What is that?" "How should we design it?" "What is it all about?"

There are issues in the design of collaboration, the design of work, and the design of relationships under corporate umbrellas. We will have several significant presentations to bring us up-to-date on what some of the design issues affecting that kind of collaboration are. There are issues in terms of the design of work, especially in an era when commuting is not getting into a car, but pushing a button or lifting the lid of your laptop computer.

So design is important to business. We are seeing more and more hyphenates out there. Those of you who are familiar with Hollywood know that they traffic in hyphenates. You are nobody unless you're at least an actor, a writer, and a producer-director. And these days there's more and more discussion about the notion of the manager as designer. Henry Minsbrook, a well-known management theorist, wrote an article fairly recently on the subject of the manager as a designer. As I have met CEOs of companies and entrepreneurs, one of whom, Hatim Tyabji, you will meet tomorrow, I've found that what they think about is designing an organization and environment—a mode of collaboration in which extraordinary things can take place.

Design is also central to business because businesses everywhere are very concerned about the issue of creativity, not just in the narrow sense of new products and services, but also in terms of new ways of doing things and new ways of understanding themselves as they hurtle forward into the future. Organizations all around the world are groping for a system, or method, for handling creativity. Twenty-five years ago, if you went to a corporation and said, "What is your system for strategic planning?" the response would have been, "Huh?" Well, these days there are systems for strategic planning. If you went to a similar corporation today and said, "What is your system for creativity?" the answer would probably be, "Huh?"

So design is important to business. I think that design is the Trojan horse for bringing a renewed spirit of creativity into the corporation. Designers know how to tap into creativity. They understand methodologies for using creativity, and so there ought to be a tremendous sense of collegiality and connection there. This makes the discontinuity between design and business all the more puzzling.

From the design side, I feel that designers have an opportunity to think differently and more broadly about business issues along the lines that I have discussed, and especially in this era of massive changes in technology. One element of this conference has been the injection of a certain amount of technology into the way we collaborate around the conference themes. When I talk about technology, it is not about debating whether it's better to use a pencil or a computer to do graphic design; it is really much more profound. It's about understanding the environment into which we are all entering as we move forward towards the future.

Anyone in this room who says they understand the digital revolution— "Oh, that stuff"—is simply wrong. Because none of us understand it; none of us know where we are going. There are too many unprecedented aspects to this new set of developments and many new kinds of design challenges. How do you design for a virtual city that exists only in a computer server somewhere in the city of Amsterdam, which in fact has a Web site called the "Virtual City"? People have been married in the Virtual City. Hundreds of thousands of tourists have visited the Virtual City. But it begs the question: How do you design for that?

How do you design for a market*space*, to use some language from one of my colleagues at Harvard Business School, John Sviokla, as opposed to a market*place*, especially in an environment in which commercial Web sites on the Internet are growing at a rate of 10 percent a week? Who understands that? Who understands what the opportunities or the implications of that are?

Parenthetically, how many of you have a personal Web page? A few. And how many of you have a corporate Web page? Even fewer. When those questions are posed in a couple of years, in IDCA 2000 let's say, most of our hands will be up in the air, and that will mark the dramatic kinds of changes that are going on in relation to this technology.

In talking with designers, I have also heard extensive concerns. As I boil it down, I would say that there is a set of concerns that has to do with whether design is shrinking. Ricky Wurman likes to describe this as "the fear of design becoming mascara": that it simply becomes

decoration. It becomes some kind of corporate short-order trick—
creativity to order. Whip it up. We need a designer. Let's rubberize the
handles of our power tools so that when we drop them on our feet we
won't break our toes.

I would argue something different. I would argue that design's domain is
expanding dramatically. It is good news and it is bad news. And it's a
big challenge. But I would argue that design is central to what organizations need to do.

To give you a preview of a position paper that Milton Glaser has prepared for the roundtable discussion, its title is "The War Is Over." By
implication what he is suggesting is that business has won. I would
suggest that design has won. The case for design no longer needs to be
made. It is central to organizational purpose. It is what people do. It is
about creativity. But it begs the question of what designers are going to
do about it.

The issue is not design, but designers, especially in this new environment with its blurring of conceptual boundaries—where architecture
can refer not only to physical buildings but to information, or software
or corporate structure; and where the breakdown of the traditional
relationship between a designer and a corporation will only accelerate
as time goes by and new pressures come to bear on that relationship.
This suggests that many other forms and business models may be more
current or appropriate—whether it is the designer as entrepreneur, or
whether it is different kinds of alliances between designers and corporations, or whether it is radically different kinds of design firms that are
going to exhibit the kind of strategic hardiness that will enable them not
only to survive in the twenty-first century, but to thrive.

This is a report from my excursion over the last year as I have been
preparing for this conference. There is a perplexing discontinuity
between these two worlds, which need so much to be partners. I think
that is our opportunity over the next few days. We have the opportunity
to explore and reinvent or reconfirm this relationship given the current
realities and environment in which we are operating.

13

To do that we have a lot of help and extraordinary resources. We have a terrific group of speakers who have come from far and wide to share their experiences in a broad range of disciplines, while always focusing back on these central questions. The virtual conference is an extremely important part of what we are here to do. In the Paepcke Gallery we have six computer terminals linked by a local area network that are running a special version of Lotus Notes given to us by the Lotus Institute's Learning Network. That is going to be our electronic infrastructure for sharing ideas, communicating with each other, reacting to the speeches, and reacting to the comments that others have made about a particular topic.

We also have our satellite feed to Japan, which will provide us with a constant reminder that design is something global as well as local and of the collaboration that can come through technology.

So my hopes for this conference are that we are going to get a chance to examine this relationship between design and business critically and collaboratively. None of us has all of the answers, but the answers reside in this room and in our ability to interact with one another over the next few days.

Initially, when I pictured this conference, I had in mind the metaphor that we were preparing for a Broadway show that was going to open and close very quickly. It was going to have one performance—kind of a poignant thought. But I subsequently revised my idea of what the appropriate metaphor for this event should be. It is not so much about Broadway; it is really more about jazz, more about the art of improvisation.

We've selected sheet music. We've got speakers. They have topics. They know what they are going to be talking about. We know where the improvisation is going to happen—it's going to happen in this beautiful meeting hall. We've got some agreement about what instruments we're going to use and how fast we are going to play, but we really have no idea about what the outcome of the conference is going to be. We are going to be creating on-line together, using some very powerful tools including our computer network.

And if I were to reflect on what makes for great jazz, I would say it's a blend of expertise—to get to play your instrument well—and being able to be a beginner, being able to take out a blank piece of paper and start from zero in trying to understand things. It's a balance between listening and playing; you have to have those in sync as well. And it's the balance between freedom, which is what improvisation is about, and discipline. We want our conversations to be free, and we want them to be grounded. We are a big jazz band here. We have got some great instruments; we've got some great tools; and we need to collaborate to find answers to the questions that have been posed for this conference.

Design Mindfulness
Tom Peters

It is a pleasure to keynote this forty-fifth International Design Conference in Aspen. That's particularly so, because back in my distant past I harbor a secret. I began my university studies at Cornell in architecture. Truth is, I dropped out after just one semester. I've always imagined I made the wrong decision. I have always been a frustrated architect and designer; and now to be invited here to keynote this event—it's little short of mindblowing.

While John Kao and I largely agree, I'm afraid I'm a little less optimistic than he is. In short, I think the world is awash in look-alike, taste-alike, feel-alike products and services, and that the ability of design to transform an institution is not yet as widely accepted by top management as he suggests. To make that point, I want to speak tonight about institutions that do get it. Those that do, though there are no perfect models, exhibit a series of thirty-five traits that add up to what I call *design mindfulness.*

1. Design mindfulness is a core competence, which becomes effective if (and only if) it becomes a culture of design. It's not just that design is important. It certainly isn't that design is just styling. It's that design can become a way of life, an avenue for competitive advantage, and perhaps—holy of holies—sustainable competitive advantage. But *way of life* is the key. That's why I've used the term "culture of design." If design is something thought of after the fact, or even before the fact but in a superficial fashion, then you don't get it. And "it" is the word. Design sensitivity—of the sort that marks Boots the Chemist in the U.K., for instance—is the name of the game. So, think core competence.

2. Design mindfulness/a culture of design is arguably the number one antidote to the commoditization of products and services. What have we been up to for these last ten, madcap years? We have TQMed everything. We have made the customer king, created the learning organization, installed the virtual organization. We have re-engineered the kitchen sink and empowered the plumber. We have sped up product development immeasurably.

Still, or more than ever, we are awash in products and services that are boring. Products and services that look like everybody else's overwhelming arrays of products and services.

In the last decade there has been, literally, a tenfold increase in the number of new products introduced to U.S. retail shelves. Yet the chief executive officer of Wal-Mart, David Glass, recently felt compelled to say to me, "There is an absolute dearth of new and exciting fashion-forward products. . . . It's become a replacement business." He added, where's the equivalent of the early Sony Walkman, the early VCR, the early microwave oven—products that sucked customers, by the millions, off of their couches and injected them into Wal-Mart stores?

A jillion new products arriving faster than ever; and the quality of almost all of them these days is good. But do they sparkle, suck the customer for financial services software or toothpaste off the couch? Rarely.

A while back I attended a seminar to mark the launch of *Collision* by Mary Ann Keller, who is generally acclaimed as the top Wall Street analyst following the world automotive industry. Her book was about the three giants in each leg of the Triad: General Motors in North America, Volkswagen in Europe, and Toyota in Asia.

To my delight, GM topped that list (though Keller quickly added that it was a matter of least worst, not best; but given GM's position in the recent past, we should take what we can get). More surprising, Toyota was at the bottom of the heap.

I asked a few questions: "What about Toyota's vaunted quality?"

"Just fine," Keller said.

"And its lean production system?"

"Just fine," came the retort once more.

"But then what's the problem?" I asked.

"The product, it just isn't very interesting," came the ready answer.

As is my habit, I immediately transformed her response into a 35mm slide, which I used for the first time during a seminar in Australia a while later, realizing that sitting in the front row, amidst two thousand or so people, was the chairman of Toyota Australia. So I trashed the company and off we went to lunch. Yes, I was seated at a table with about ten other people, including the Toyota honcho.

He said Keller was right to point to Toyota's problems with bureaucracy, etc. But he added that he thought she got the specifics wrong: There wasn't a major product problem.

I shut up, because I'm no expert. But then, maybe fifteen minutes into the lunch, in response to nothing in particular, he turned to me and said that when he's cruising the streets of Sydney or Melbourne, sometimes he can't tell the Toyotas from the Hondas and the Nissans.

I just about dropped my teeth. This was a person who only moments before told me that there was no product problem? Then, "can't tell them from"! Get serious!

Truth is, this year has had a pattern to it, and one that hasn't been that much fun. I've been called in by office furniture makers, computer companies, software companies, big banks, small banks, big insurance companies, training companies, Big Six accountancy firms, and engineering services firms. Each one has seemingly had, when you boil it down, the exact same problem: "New competitors springing up all over; new products, with world-class quality, hitting the marketplace at a record pace; customers tightening their purse strings; distributors putting on the squeeze; margins going to hell; *my product/service is becoming commoditized!*"

"Becoming a commodity." If I hear it again, I think I'll throw up. Me too. Me too. Me too. Me too-ism galore. For heaven's sake, just a few weeks ago I spent a day on the floor of the very prestigious International Contemporary Furniture Fair at the Javits Center in Manhattan. Ninety-eight percent of the stuff that I saw there looked just like everything else. It was insipid. Uninspiring.

Which, of course, translates into uninspiring retailers. At each of the big fairs, the owner of a successful specialty shop wrote, "Crate and Barrel buyers are hard on the heels of Williams-Sonoma buyers who are hard on the heels of Pier 1 Imports buyers. This means that what is available to the American consumer looks very much alike from one store to the next." Amen. And, "Sony," said *Tokyo Business* in its August 1994 issue, "has changed from launching breakthrough products to launching price offensives on 'me too' products."

"They are selling us tired clothes," said Benetton's top Italian franchisee about that oh-so-recently-proud firm.

Inevitable commoditization in the face of a global explosion of new products and competitors. That's it. Right? Never!

I remember the day so well, near the end of February in 1994. My *Fortune* arrived, the annual issue featuring America's most admired companies. The first few years, a no-brainer. IBM wins going away. Then that firm, in effect, imploded. The replacement on *Fortune*'s list? No surprise, Merck. And then came the 1994 issue, and there on the cover, the pride of Worcester, Ohio—that's right—Rubbermaid!

Rubbermaid, maker of rubber and plastic doodads. About four hundred new products each year, over one per day! Voted the most innovative company in America by one magazine. Tied for first place with Microsoft in number of industrial-design awards won in a prestigious 1994 contest. Ninety-five percent of Americans recognize the brand name, which puts Rubbermaid in a league with Coke and Disney (and just about at the top of that league).

I love it. I simply, thoroughly, purely love it. Because if you can differentiate Rubbermaid's "stuff," then the rest of us have no excuse, no excuse whatsoever. If they can do it with that (holds up a handful of Rubbermaid products), then all of us who speak of "inevitable commoditization" should hang our heads in shame. Or better yet, quit.

All this is a long-winded prelude to my larger point: We're gathered here together to talk about design or, writ large, *design mindfulness* in my lingo. And I believe that design mindfulness is arguably—no, make that *clearly*—the number one route of attack for standing out, for halting, and even reversing "inevitable commoditization," and for "Rubbermaiding" our consulting services and training services.

3. Design mindfulness looms ever larger as the economy grows ever softer. The end of the industrial era came, unheralded, on a day in early 1992. Microsoft, then only booking about $2 billion in revenue, watched its total stock market value surpass that of General Motors, then tallying almost $125 billion. As we speak, Disney's stock market value is greater than Ford's. Well, you get the drift.

Maybe the best certification of the new, "soft" age came a few weeks ago, when His Eminence, "Air" Jordan, was vacillating about returning to the business of bouncing basketballs. For a week he held us in suspense. And during that week, the stock market value of the five firms whose products he shills—Sara Lee, Nike, McDonald's, Quaker Oats, and General Mills—increased by $2.3 billion.

Look, it's simple. The nerds won. You and me. I spent much of my career as a consultant. We were parasites. So were the ad folks, the accountants, the lawyers, the designers—all the providers of professional services. We were pond scum, living off the honest sweat of real men's brows. No longer. Now, they, the lifters and haulers, are the new parasites. We are the engine of the new economy, the brain-based economy, the "idea-based economy," the soft economy.

There was a bit of a gold-mining boom going on in Australia when I was last there. The source? Bigger, tougher steam shovels? Taller Aussies? Hardly. Mostly, sophisticated geophysical and geological software designed by nerds, which allowed the Australians to find previously hidden ore deposits.

Research by Murikami Teruyasu at the Nomura Research Institute concludes that the Age of Agriculturalization is behind us; so, too, the Age of Industrialization. We're in the midst of the Age of Information Intensification. But it's not the last act. Coming on strong is what Teruyasu calls the Age of Creation Intensification. I think he's right, in general. Sure of it, in fact. And I know he's spot on for those of us in the advanced, fully developed countries.

Look, good products, top-quality products, come from all over the world these days—from most of China and India and Indonesia, from Malaysia and Korea and Thailand, from the Philippines and from Chile and Argentina and Brazil and Mexico. What's going to set us apart? Keep us at the top of the heap? The only answer: plunging gleefully into the new age, the Age of Creation Intensification, the Age of Soft, the Age, perhaps, of Design Mindfulness?

4. Design mindfulness is a "numerator issue." "Lean, mean . . . and then what?" is the way one executive put it. "Lean and mean does not a strategy make," is the way I put it in reply. There was a wonderful ad recently for Mercer Consultants. The tagline: "You can't shrink . . . to greatness." Marvelous!

Gary Hamel, management guru and co-author of *Competing for the Future*, told the *Financial Times* that 1995 should be the "Year of the Numerator." That is, the time has come to focus on the top line. The revenue line. To be sure, most of our firms were far too fat and far too bloated. Many, perhaps most, companies still are. Nonetheless, the lean up-mean up process is merely a precursor to competitiveness. Competing effectively is about doing good things, making good things, creating good things. It is clearly a numerator issue. Design, too, is just that: a numerator issue.

5. Design mindfulness is a pervasive notion. Designer Michael Shannon, in a letter to me, said it as well: "Design is how a company looks, feels, tastes, wears, rides . . . what the company is that customers care about."

I like that a lot. It's about wholes, about sensations, about the fact that this company cares, dares to stand out, pays attention to just how "neat" its products are.

6. Design mindfulness is a matter of character. In *Design Management Journal*, designers Peter Laundy and Susan Thornton Rogers attack the "marketing warfare" approach to business taken by gurus Jack Trout and Al Reis. Marketing, they tell us, is angling for advantage, positioning.

No, no, never! Not if you've got good sense, Laundy and Rogers almost snarl. They write,

> Position is a concept that lacks passion, dealing with cognitive place but not emotional resonance. It also lacks conviction: The point [of positioning] is to find the open spot rather than to follow one's thoughts and feelings. A company that allows itself to be defined by competitors [in a battle for best position] ends up without intrinsic identity. Character, an inner-centered viewpoint, is concerned more with who a company is than where it is in relation to others.

To be sure, such internal focus has gotten many a company in trouble. Deep trouble. On the other hand, attending to issues of character, surprisingly, may be the best (only?) way to break out of the box. "Paradoxically," Laundy and Rogers conclude,

> in seeing the world in a less competition-centered light, the Character-Expressing Company is more likely to reinvent the market instead of getting sucked into playing the game in a similar way as its competitors.

The notion of design as character is exciting. So, too, the idea of character, per se, as core competence: attending to (obsessing about) what feels distinct and honest.

7. Design mindfulness flows from leaders who love stuff (and vice versa). Testifying before a congressional committee in Washington a while back, I was joined by two CEOs, one from a middle-sized company, one from a giant. The chief of the smaller company, which has an exceptional reputation for zippy products, cornered me for a few minutes—to talk about the trade issues before us that day? Never! Thirty seconds into the conversation, with hardly an exchange of courtesies between us, he was going on about his product. He couldn't help himself, couldn't stop, handed me product-literature, invited me to visit, burbled on and on and on.

Subsequently, I must have chatted for ten minutes with Big Co.'s Big Boss. I'm not exaggerating: At the end of the ten minutes, I had no more idea of what his company produced than I had when I walked into the room. No mention of product. None. None whatsoever. Did he sell accounting services? Or rocket ships? To this day, I don't know. (I do know, of course; his firm is a household name. But I don't know based on our conversation.)

The *Economist* insulted us! The statement is, well, so simplistic. The magazine reviewed a study of innovation that focused on superior performers, labeled "product juggernauts." Just what can we say about a product juggernaut? "The hallmark of a 'product juggernaut'," the *Economist* wrote, "is an unnatural obsession with what it produces."

That's it, "unnatural obsession"! Sadly, as my recap of the two conversations suggested, it often is "unnatural" for Big Cheese to be intimately engaged with their products.

But not so if the company is Rubbermaid. And if the boss is CEO Wolf Schmitt. "I'm probably the biggest pain to our people in sending out notes with ideas and clippings," he says.

> I'm probably on every mailing list in the world because I buy a lot of the products I find in catalogues. I send them to people to stimulate new ideas. It's easy. It's fun. People then recognize that you really want this sort of activity going on. I enjoy new products. I like color. And I think people understand that. As a result, it's contagious. Maybe that's what it's all about. You sort of inoculate everybody with that desire to do those things.

Got that? The secret to success at Rubbermaid: The boss clips pictures from catalogues and sends them out to everyone. It is just about that

23

"simple." Except, underneath, is that abiding passion, that love, that unnatural obsession with the product.

8. Design mindfulness is a passion for artistic expression at age eighteen or sixty-eight. Design mindfulness is the stellar designer with a bushel of awards. It's also an eighteen-year-old waitress with attitude. The restaurant's not that great, but she arrives for her shift full of vim. For the next several hours, she "owns" her five tables. She'll experience, on average, fifteen sittings, thirty-eight guests.

Just as is true on Broadway, each sitting is a performance. Each guest is a unique opportunity. That's right. That night in the diner (or four-star restaurant) provides, for her, thirty-eight perishable, never-to-be-repeated chances to stand out, forever, in the minds of the people she serves. Or not. That's art. High art.

Or, as psychologist James Hillman put it in *Kinds of Power*, service at its best can be "superb, graceful, beautiful, divine, marvelous, wonderful . . . an aesthetic gesture." That, I'd add, is the epitome of design mindfulness.

9. Design mindfulness can be thoroughly populist. I've learned at least one lesson in the last ten years. When I'm on my farm in Vermont, duct tape is the answer for nearly everything. No, make that everything! Duct tape. Rubbermaid products. Ziplocs (oh, the glory of Ziplocs). Post-it Notes.

I've read the debates, which I think are mostly foolish pedantic exercises, about the role of elitism in art. Sure I think elitism has a role. I've got no problem with the multi-million dollar pieces I see in the Museum of Modern Art or the Metropolitan. I love many of them, in fact; and I clearly understand that they end up—twenty years later—influencing many aspects of everyday life.

But design mindfulness, and art for that matter, can also be expressed by Wolf Schmitt's followers at Rubbermaid. Design mindfulness for the masses is no pipe dream.

10. Via design mindfulness you can wildly differentiate a 79¢ product as well as a $79,000 product. I used to be a snob. I've always favored high-differentiation strategies, as opposed to low-cost strategies. High differentiation to me meant Rolex, Mercedes, and so on. But I was wrong. And Rubbermaid reminded me of that. Major differentiation, wild differentiation, amazing differentiation can be had for less than a buck. That is, "high-end" (distinguished) does not equal high price. It's a matter of attitude, or call it design mindfulness.

11. There is a vast, mass market for good design. In the late 1940s, the designer George Nelson fought conventional wisdom and brought the idea of a big market for good design to Herman Miller. He was right, then and now.

I've had it up to my eyebrows. I'm sick to death of, in particular, the notion that "Americans don't have good taste," "Americans aren't attracted to well-designed items." I'm sure there are numerous Americans (and Japanese and Germans and Swiss and Swedes and Malaysians) who don't give a shit about design. But, given the choice between something "neat" and something that is decidedly "not neat," assuming rough price parity (and remember I just said that superlative design is possible in seventy-nine-cent products—a point that I passionately believe), the "average Joe" and "average Jane" will opt, I confidently predict, for the neat. And if you don't buy that? Well, get a life!

12. Design mindfulness is as potent a tool for small companies as for large ones, for high-tech firms as for low-tech, for service enterprises as for manufacturers. The tragedy is that most small companies, in particular, don't get it. (This is one big reason I'm seriously considering writing a book about design.) They don't understand that design clearly is the principal path to standing out from the crowd. The British Chartered Society of Designers' booklet, *Managing Design to Sharpen Effectiveness*, features an unlikely company, Thrislington Sales, a manufacturer of toilet cubicles. The firm's policy has become "design excellence in all aspects of its [business]." And design per se "has transformed its products, literature and working environment to the point where all are coordinated to present a coherent design vision both to customer specifiers and to the world at large."

Thrislington Sales had suffered with a me-too product in a decidedly commodity market. The managing director once attempted to peddle his wares to a big potential customer in the Netherlands. "He had not even had the chance to finish unpacking it from the back seat of his car," the Society's booklet comments, "before he was told he was wasting his time." Following design-led total transformation of the company, the story is different today. The toilet-cubicle firm has become an international powerhouse, and the formerly dismissive Dutch are buying by the truckload.

An even more homely story is recounted by author and entrepreneur Harvey Mackay. The topic: a cab ride from Manhattan to La Guardia:

> First, this driver gave me a paper that said, "Hi, my name is Walter. I'm your driver. I'm going to get you there safely, on time, in a courteous fashion." A mission statement from a cab driver! Then he holds up a *New York Times* and a *USA Today* and asks would I like them? So I took them. . . . He then offers a nice little fruit basket with snack foods. Next he asks, "Would you prefer hard rock or classical music?" He has four channels.

Mackay goes on, listing other features such as a cellular phone for the passenger's use. Then he concludes, "You know what? This man makes $12,000–$14,000 extra a year in tips. You should have seen the tip I gave him. Incredible."

Is Mackay's tale about design mindfulness? By my lights, the answer is yes. It is the total transformation of the ultimate "commodity" (a Manhattan cab ride!) into something special, something beautiful, something wildly different from the norm. And talk about small business!

13. Design mindfulness is a big deal. Historically design has time and again transformed entire industries. What's the big deal? People keep telling me, "No one appreciates design." I understand that not enough companies, large or small, service or manufacturing, appreciate design. But design-as-big-deal is not a new idea.

I devoured William Leach's *Land of Desire: Merchants, Power and the*

Rise of a New American Culture. It describes the transformation of retailing at the end of the last century and in the first decade or so of this one. The centerpiece, implicitly: design. Such things as the use of color, and, especially, the introduction of glass display windows literally transformed an entire, monster industry. Actually it's more than that: design, per se, created modern retailing.

Or consider my favorite book of the moment: Thomas Hines' *The Total Package: The Evolution and Secret Meanings of Boxes, Bottles, Cans and Tubes*:

> During the thirty minutes you spend on an average trip to the supermar-ket, about 30,000 different products vie to win your attention and ultimately to make you believe in their promise. When the door opens, automatically, before you, you enter an arena where your emotions are in play, and a walk down the aisle is an exercise in self-definition. Are you a good parent, a good provider? Do you have time to do all you think you should, and would you be interested in a short cut? Are you worried about your health and that of those you love? Do you care about the environment? Do you appreciate the finer things in life? Is your life what you would like it to be? Are you enjoying what you've accom-plished? Wouldn't you really like something chocolate?

And all of this, Hines argues persuasively, is virtually determined by packaging. That is, the role of design/design mindfulness, writ large, is a centerpiece of society. It's a big deal, mate.

14. Design mindfulness applies to the accounting and purchasing and logistics and training departments as well as to engineering, R&D, marketing, and design itself. Design and accounting? Of course! It's not just that there are few more beautiful things than a P&L awash in P. It's that accounting is about the flow of information. It can be performed elegantly. It can be performed with the beauty of perfect simplicity. Or it can be complex, messy, and ugly. It can be helpful. Or it can be obscure.

Take something as "simple" as the design of a form. Most of us trash the IRS. (Who the hell enjoys paying taxes?) And, to be sure, the agency has lots of room for improvement. (Just ask almost anybody who works there.) On the other hand, the IRS is aware of how much the tiniest difference in a form can make to the number of errors you or I will make in filling it out. Would that all of our accounting departments paid as much attention as the IRS does to the form and format of the presentation of information!

I know some artists (all too many, to be truthful) who insist that art starts and stops with the application of paint to canvas. Utter rubbish! Art in the presentation of a training course can be as potent a force as the application of some medium to canvas, paper, or whatever. That is, design mindfulness deserves its big spot in the sun in every department in the organization. No exceptions.

15. Design mindfulness is abetted by the presence of great designers—on the payroll, hanging about, and so forth. Okay, okay, I just said that design merits a prominent place in the accounting department. And I believe it. But just as Michael Jordan redefined basketball, so, too, a single designer can transform an organization. Witness, for example, George Nelson at Herman Miller.

This is the Age of Soft, Softer, Softest. The Age of Talent. The Age of Brainware. Nathan Myhrvold, head of research at Microsoft, says a top programmer is a thousand times more productive than an average programmer. The Age of Big Soft must be the Age of Big Talent. That doesn't mean a "famous" designer must be on the payroll. But it does argue for the insistent, perhaps even desperate pursuit of genuinely fabulous design talent.

And then—and I'm not saying this to get applause at a large gathering of designers—pay that talent. Jordan gets paid more than his coach, Phil Jackson. And, in my mind, in the Age of Talent, top chefs should be paid more than the boss of the restaurant, top designers paid more than managers of design and, upon occasion, the president of the company.

Talent. Talent does count. I'm all for participative management, getting everybody involved, using everyone's creativity (and I happen to believe that everybody is pretty damned creative); but there is a difference between a superstar and a non-star. Recognize it. Hire it. (Or contract with it.) Pay it.

16. Design mindfulness is reflected in formal (organization/incentives) as well as informal arrangements. The late management expert Chris Lorenz's influential *The Design Dimension* mostly focuses on the formal attributes of design—e.g., its place on the organizational chart. He traces the movement of designers to the "head table" at Ford, the presence of a vice-chairman's position for design at Sony.

Design may be an intangible, but if the designer is not even invited to the table, then that all-important intangible is going to have a difficult time making itself known.

Pay a lot of attention to design's formal position on the organizational chart, on project teams; exactly when the designer is invited to be part of the team and so on.

17. Design mindfulness translates into playfulness and an abiding appreciation of the role of emotion in corporate life. I love business. I admit it. Sure a lot of business is drab and dreary. (So, too, every other pursuit in life.) But business at its best is, well, great—creative as hell. Tapping the highest potential of large numbers of people. Transforming whole aspects of the way human beings live.

"Business is a creative activity, involving inspired hunches and leaps of faith," said Ludwig von Mises, de facto founder of the Austrian school of free-market economics. "Creative activity . . . inspired hunches . . . leaps of faith." Ah!

Professor Robert Peterson of the University of Texas did study after study in which he could find no link between customer satisfaction and repeat business. Huh?

Then he changed the scale he was using. Instead of one that went from "dislike" to "like," he used a scale that went from "hate" to "love." With the new measure, he got the correlation.

THE NEW BUSINESS OF DESIGN

People must have an emotional attachment to the product or service to make them want to come back, says Peterson. And we're not just talking pizza and lipstick. The late Paul Sherlock was a renowned product developer and marketer at Raychem, the high-tech materials company. In his important book, *Rethinking Business to Business Marketing*, Sherlock said the two most important attributes a sophisticated product can have are "glow" and "tingle."

You're making a presentation. You see, said Sherlock, the eyes of the folks on the other side of the table light up. You know that you've got them. Sure, they'll have to produce some kind of a sophisticated, written rationale to justify buying your product. But whether that product is a $100 million a year information-systems consulting contract or pens and pencils, the "glow" and "tingle" are, by and large, decisive.

Emotion: in our business schools and in our businesses, we tend to downplay the role of passion, engagement, and emotion. It's a mistake. It's always been a mistake. But now it's a tragic mistake. With the world brimming with new competitors and look-alike products, how do you stand out? How do we maintain our relatively high wages? The answer is products and services that are distinct. Special. Products that glow and tingle and make you fall in love. Products, that is, which are *emotional*. The path toward high emotional content? Near the top of the chart: design mindfulness.

18. Design mindfulness and daring-to-be-different are handmaidens. "We are crazy," Canon president Hajime Mitari told *Forbes*. "We should do something when people say it's crazy. If people say something is 'good,' it means someone else is already doing it." Those words should be tacked behind the desk of everyone in this tent or program it onto your screen saver.

The pursuit of crazy. An honorable pursuit. The only pursuit. But it's the point about "good" that really gets me. Many times, as a consultant for McKinsey, I was called in by a tired company to present a strategy. After a lot of work, we came up with something that passed my "neat test," and, according to the response, the client's neat test as well.

Having gotten excited about the special opportunity, the next words out of his mouth, more often than not, were, "Can you find me three examples of people who are already doing this?" To say he didn't get it is an understatement—a gross, sickening understatement.

Or consider this corroborating evidence from *The Starship and the Canoe*, a book about the physicist Freeman Dyson:

> Freeman Dyson has expressed some thoughts on craziness. In a *Scientific American* article called "Innovation and Physics," he began by quoting Niels Bohr. Bohr had been in attendance at a lecture in which Wolfgang Pauli proposed a new theory of elementary particles. Pauli came under heavy criticism, which Bohr summed up for him: "We are all agreed that your theory is crazy. The question which divides us is whether it is crazy enough to have a chance of being correct. My own feeling is that it is not crazy enough." To that Freeman added: "When the great innovation appears, it will almost certainly be in a muddled, incomplete and confusing form. To the discoverer himself it will only be half understood; to everybody else it will be a mystery. For any speculation which does not at first glance look crazy, there is no hope."

Dare to be different. Shockingly, that's still a strange idea in 1995. The same people who call me to bitch about "inevitable commoditization" apparently wouldn't understand what Mitari of Canon was saying, nor Freeman Dyson. Or if they do understand it intellectually, somehow they can't develop the moxie to do something about it in the real world.

19. Design mindfulness strikes a blow against boredom. In *The Death and Life of American Cities* the philosopher Jane Jacobs decried sterile, urban planning of the 1950s variety. Cities suffered, she said, from the "Great Blight of Dullness." Where was that exciting mixture of residential and business, low-rent and high-rent? It was precisely such a mix, which Jacobs felicitously labeled "Exuberant Variety," that made all the difference. Oh my, oh my: Exuberant Variety! And do those words ever deserve the capital letters Jacobs invariably gave them.

Screw things up. Shake things up. Kevin Kelly, editor of *Wired* and author of *Out of Control: The Rise of Neobiological Civilization*, says,

"Equilibrium is death . . . seek constant disequilibrium." The boss gets paid to screw things up. The designer gets paid to screw things up. Design mindfulness is being a disrupter.

But the accolade for the best description goes to Ludwig Wittgenstein: "If people never did silly things, nothing intelligent would ever get done." Hats off to silliness. (That is, if you're in pursuit of the intelligent.)

And the last word on this topic. Honors go to Tony Hendra, writing in a scholarly economic journal, *GQ*:

> What did the Fifties, that economic Eden we're all trying to get back to have that we don't have? A Republican president? Nah. Lower taxes? Nah. It's much simpler than that: People drank at lunch.
>
> Our economy revolves around lunch. Lunch, for the early jogging, hard-charger, is the first meal of the day. Appetites sharpen; greed is at its peak. People make deals at lunch because they're hungry. But mere deals are not enough. What boosts economies into orbit are insane ideas—inventing an oven that will cook things in minutes . . . developing a pill to prevent pregnancy . . . the kind of ideas that seem demented if you're drinking fruit juice but make complete sense if the juice has been fermented.
>
> For money to be made, someone has to say to someone else "Yes." And for lots of money to be made, someone has to scream "Yes! Yes! Yes!," whoop, holler, high-five, clink glasses and throw bread at the other tables. What made the fifties' economy wasn't good old family values. It was the three-martini lunch.

Forget "good." Love "crazy." Stomp out the Great Blight of Dullness. Wipe out boredom. Disrupt the status quo. All hail Exuberant Variety. Pursue the silly and throw bread at the other tables. That's the idea.

20. You want design mindfulness? Then hire (and reward and promote) design mindfulness. Recruitment equals strategy. That's obvious. But not so obvious when you watch it being practiced. Watch it being honored in the breech.

If we want to fight the Great Blight of Dullness, then we must pursue, not eschew, the other-than-dull. Hunt, perhaps, for Dudes with Poetry.

Pay close attention to these observations from Stanley Bing, writing in *Esquire* about an encounter with a job candidate that went nowhere:

> He comes in and seats himself carefully on the edge of my guest chair. He is staring at the toys on my desk, trying to suppress the realization that I am an infantile nit whose job he could probably do much better. . . . Of course he does not play with the toys. . . . He looks out of my window instead. "Nice view," he says rather perfunctorily, but he does not say, "Wow!"—which is what my view of the canyons and spires of high-mercantile capitalism deserves. . . .
>
> "I'm looking for an entry-level position in public relations. Maybe corporate marketing, if I get lucky," he says.
>
> "Really?" I say. "Like, out of the entire realm of human possibility, that's what you want to be doing?" I'm sorry. He's really starting to tweeze my bumpus. What twenty-four-year-old really and truly wants to be in corporate marketing, for God's sake?
>
> I look him over as he burbles on about targeting demos or retrofitting corporate superstructures or some frigging thing like that. The guy makes me want to stand up on my desk and yell, "Booga-booga!" Instead, I say, "Didn't you ever want to be a rock musician or a forest ranger or anything?" He looks at me like I have a banana peel on the end of my nose. It's quite clear to me that, since he was in high school, he's been preparing to be a . . . communicator. That's actually what he says.
>
> Screw it. There's no poetry in this dude. No soul. No surf or wind or whale bone in his eye. He's . . . desiccated. He makes me sad. I kick him out of my office.

Three cheers for Stanley Bing! And three more cheers for Cheryl Womack, boss of the fast-growing, specialty insurance-products company, VCW, of Kansas City. Her secret to success, she says, comes in the hiring. "We look for passion, flexibility, excitement," Womack asserts. I like that. Says she can teach the insurance issues. Can't teach passion, flexibility, excitement. Hence, recruit it. And when you get it, and watch it explode, then promote it. Promote it fast. And pay it well.

Design mindfulness is the search for passion, flexibility, excitement, and Dudes with Poetry.

21. Design mindfulness is perpetual curiosity. "Twelve is about Ben's real age," wrote Roger Rosenblatt of his pal, the legendary *Washington Post* editor Ben Bradlee. Bradlee himself owned up to "compulsive spontaneity" and "advanced immaturity."

Don't grow up! If it is, indeed, the Age of Creation Intensification (remember the analysis from Nomura Research Institute), then the pursuit of perpetual curiosity is the highest art, which means a willingness, from bottom to top in the organization, to support advanced immaturity, compulsive spontaneity, and people who work at never getting beyond age twelve. Hardly a normal recruiting routine, right?

I said I love business. I love businesspeople, too. Maybe not the average bureaucrat. (Even though he or she is probably a good parent, etc.) But I love Anita Roddick, Body Shop founder; and Herb Kelleher of Southwest Airlines; and Luciano Benetton; and Virgin Group's Richard Branson.

The ones I love, the Kellehers (et al.) of the world, are mad as hatters. Al Neuharth with his "stupid" idea, *USA Today*. Ted Turner with his even nuttier CNN. You know the type.

These folks prove to me that economics is not a "dismal science" in actual practice, that business is not uninspiring. And they prove to me that perpetual curiosity, spunk, spirit, and a touch of madness are the essence of long-term success and yet another manifestation, in my book, of design mindfulness writ large.

22. Design mindfulness means that all decisions must pass the DMT, or *Design Mindfulness Test*. In companies where quality reigns supreme, every issue, related or unrelated to quality per se, must go through the quality sieve. Does this fuel our obsession with world-class quality? That's the question any discussion hinges on, whether the so-called topic is family leave policy or a minor change of accounting practices.

Let's pursue the analog. In this, the Age of Soft, Softer, Softest; the Age of Creation Intensification, everything we talk about in the firm must pass the Design Mindfulness Test. How does this tweak to recruiting practice enhance our commitment to spunk, spirit, verve, and *joie de vivre*?

23. Design mindfulness and diversity are identical twins. I'll be called in by top managers at a multi-divisional company with products suffering from a terminal case of the blahs. They want me to keynote a strategic "offsite retreat." I walk into the meeting hall, in West Palm Beach or Palm Springs, at 8:00 A.M. I've digested page after page of turgid analysis. My head is swimming. But now I spot the problem in a moment. (A few seconds, actually.)

A hundred people are there, representing ten divisions in a $3 billion company. Ninety-eight are old white males. Average age: forty-eight. Standard deviation of birth date: nine hours. (Just kidding . . . barely.) And while there may be a bit of diversity in dress in the executive suite, at this offsite retreat, 95 percent of the old-white-guys are wearing CPLGGPs: classic polyester, lime-green golf pants.

And they wonder why their products are not totally scintillating—aren't attracting the burgeoning Hispanic market, the African-American market, women!

If you're in favor of diversity for the right reasons, God bless you. But if you're into greed, pure greed and nothing but greed, well, then, sign up for diversity. Find me a car company with only one woman among the seventeen members of the executive committee, and I will have gone a long way toward explaining why you're not doing a great job serving women—who make up 50 percent or so of the primary decision-makers in modern car-buying situations.

All creativity, virtually everyone who has studied the subject agrees, comes from the juxtaposition of "odd stuff," stuff that doesn't fit together comfortably. What could be a better misfit than African-Americans and Asians and Hispanics and whites? Young and old? People from poor backgrounds and people from rich backgrounds? Tall people and short people? Fat people and skinny people? Gays and straights? Men and women? That's what diversity is all about. It's always been a good idea, as far as I'm concerned. But now it's an imperative, an economic imperative in the Age of Creation Intensification. Yes, you can definitely call diversity a design-mindfulness imperative.

24. Design mindfulness is about usability, manufacturability, and commercial viability. Design is what makes you shout "Holy smokes!" And that's important. Design is beauty. And that's important. But I also want to emphasize, stealing from premier designer David Kelly of IDEO, that design (and design mindfulness) is usability, manufacturability, and commercial viability as much as (or, cumulatively, more than) "style" attributes. That's why Kelly's famous product-design program at Stanford emphasizes engineering and business as much as art.

Design mindfulness is multi-dimensional—period.

25. Design mindfulness is a thirst for personal renewal. These are taxing days. We have to run faster and faster and faster. And yet a premium, as never before, is being placed on creativity. It's the ultimate paradox: Run faster than ever; but take longer breaks or whatever it takes to get genuinely retooled. (I stole this idea from Rubbermaid's CEO, a speed demon who champions creative reflection as a national imperative.) It adds up to what I call the case for "uppercase R" Renewal.

"Lowercase r" renewal? That's okay. Attend a course here, a course there, take three-week vacations every summer. But what are you doing to really blast the cobwebs out? I like to think of myself as a humanist, but in this (one) instance I'll give a nod to the accountants: I think we need to consider Renewal in terms of what I call the Depreciating Asset/Renewal Investment Model. The half-life of our spunk, as well as our particular knowledge, is only a few years. Four? Five? Six? Surely less than ten. We have to coldly think about ourselves as losing 20 or

25 (or 30 or 35) percent of our marketplace value each year. If the topic were factories, we'd know we've got to put enough money aside to replace what we're losing! When it comes to you and me, the story must necessarily be the same.

Yet how many of us are conscientiously paying attention to our Big Renewal programs with the same intensity we apply to choosing a vacation spot, or selecting a new car? Not that many.

In fact, the sum of all personal renewal plans is equal to nothing less than the strategic vitality of the firm and it's design-mindfulness score. As a boss, if you're not paying attention to individual personal renewal plans (starting, natch, with your own), then you're not interested in spunk, curiosity, or design mindfulness.

26. I know it—design mindfulness—when I see it, or there are limits to measurement. "What gets measured gets done" is a favorite saying of practicing managers. And I champion the idea of focused measurement in general.

On the other hand, whether the issue is quality or service or design mindfulness, it is something that we "get." We know it's there or not. Trusting our gut and putting intuition and holism, rather than standard, analytic-centered reductionism, at the top of the management "tool" list is imperative.

"Quality doesn't have to be defined," Robert Pirsig wrote in his novel *Lila* (Pirsig is best known for *Zen and the Art of Motorcycle Maintenance*). I've gotten into many a fight with the quality freaks over this sentence. They say quality is in the SPC (Statistical Process Control) charts. I say the charts are fine (essential, even) but that the quality fanatics who have created quality-fanatic companies know it when they see it—even if they like to have the charts around, too.

27. Design mindfulness is impossible to pin down and easy to spot. It's another plea to trust that gut feeling. A friend is a very successful artist. I remember the first time that I walked into her studio. I'm very conscious of workspaces and try to make my own as comfortable and energetic as I can. But when I walked into her studio I knew I'd found Nirvana. Something about it fit: fit her work, fit her, was serene, was dynamic, was controlled, was kinky—everything was "right." Nothing was "wrong." It was a mix, a jumble, crazy, and it worked. The ability to use such language ("it worked"), the willingness to bet the company on such language, that's the essence of the firm that pursues design mindfulness for strategic advantage.

I also call the phenomenon the "no electron microscope test." When a product jumps out at you well, it does just that: jumps. The first Ford Taurus. The new Neon from Chrysler. I've seen several automobile ads loaded with little arrows pointing to the "exciting" new features that differentiate it from the last (uninspiring) version. Well, if you need little arrows to tell how "fabulous," "exciting," and "stupendous" you're thingy is, you're in deep trouble. When you go into a restaurant that's great, well, it's great. You needn't lug along an electron microscope to prove it. Okay?

28. Design mindfulness flows from autocratic leadership. The chief design adviser to Boots the Chemist (Britain) talked busy top managers into getting heavily involved in the details of what had been a chaotic, fragmented approach to design.

Micromanaging. Bad stuff, says the new conventional wisdom. By and large, it is bad stuff. Folks down-the-line need lots of space to express their creativity. But when it comes to character there needs to be a strong, clear voice from the top.

It came up in a nonprofit organization I'm involved with. We'd been muddling about for years, trying to figure out what we were doing. Strategy documents lined the walls. And we were moving forward. But what, exactly, was the hook, the very special trait that would suck in widespread support?

There was a big gathering planned, aimed at getting the entire town involved. Someone suggested printing bumper stickers. Terrific! But what were the bumper stickers going to say? Delegate that to a bumper sticker or promotional committee. That was the idea some people had.

But someone leaped in—no, design of that bumper sticker, in every detail, was a board-level issue. Turns out she was right. Boiling your message down to three-to-five words that can be read by people with lousy vision from thirty-five or forty feet away, traveling seventy miles an hour, well, that causes you to figure out precisely what you're up to.

So I've come to believe that autocratic leadership, meaning direct, top-level involvement in bumper-sticker design (for example) is a key to design mindfulness.

29. Design mindfulness is abetted by ignoring your customer. "The customer is a rearview mirror," says George Colony of Forrester Research, "not a guide to the future."

Listen to the customer. Get close to the customer. That's what I preached in 1982 in *In Search of Excellence*. A good message then. A good message, for most firms, today. On the other hand, I always imagine myself part of the first focus group for Post-it Notes. The nerdy 3M scientist comes in, carrying these little packets of yellow paper. None of us in the room need it. We were happy with paper clips. Right?

Nonetheless, he enters, hands his stuff out, asks us to take one off the top. We peel it off and he begins his spiel: "Feel the back. It's glue, sticky. But, see, it's not that sticky. It's glue, but it's not 'good' glue—if you know what I mean. It sticks. But it doesn't stick all that well."

Are we all excited yet? In fact, it took a dozen years to make the Post-it idea stick (sorry!); yet rumor has it that, today, the product and its ancillaries rake in almost a billion bucks a year for giant 3M. But if 3M had depended on market research, on those focus groups for a yes or no decision, well, none of us would be using Post-its today.

In fact, it's my observation that as reliance on market research has increased, the me-too-ism score of products and services (and politicians, for that matter) has increased exponentially.

Chuck Williams, founder of the high-end cookware retailer-catalogue Williams-Sonoma, explained his secret of success. "I just bought what I liked," he said. "I never bought anything I didn't like. Fortunately a lot of people liked what I like."

"I'm not creating a game," says one of Nintendo's leading game designers. "I am in the game. The game is not for children, it is for me. It is for the adult that still has the character of a child." Both are excellent ways of saying, "Follow your bliss." Your bliss may get you into trouble (it will constantly), but pandering to customers is likely to produce a string of uninspiring products and services. "I never bought anything I didn't like" sounds to me like the basis for character, the secret of success, or at least the secret to surprising your customers.

30. If you are serious about design/design mindfulness, it will consume you. Roger Milliken of Milliken & Company eats, sleeps, and breathes quality. His firm has been called, by one important magazine, the best company in any industry in the world.

Milliken & Company has charts and graphs that depict its progress in quality. It has all the accoutrements of successful quality programs. But the essence of the abiding commitment to quality at Milliken is the obvious fact that Roger Milliken is consumed by the topic. So, too, was Bob Galvin, when he was chairman of Motorola.

Consumed. That's the idea. If design mindfulness is to be the essence of the enterprise, then leadership will eat it, sleep it, walk it, talk it. Will be consumed by it. Will buttonhole you to talk about it. Will buttonhole anybody to talk about it. Consumed, a big word.

31. Design mindfulness is strategy. Strategy is not a thirty-page—let alone a thirty-volume—set of documents. Strategy is not attack plans for cornering the market. (I've got no problem with attack plans; but they aren't strategy, they're tactics.) The essence of strategy is character, the feel of the enterprise, Michael Shannon's "look, feel, wear, ride." Design mindfulness is strategy—an encompassing way of life. It's where we've been heading all along. If design mindfulness is anything less, then the culture of design that has the power to literally transform an industry is not present.

32. Design mindfulness as core competence requires constant reinvention. The world is changing fast, absurdly fast. Throwing baby parts out along with the bath water is imperative. I know it's an ugly, scary image. That's precisely my point. Blowing up core competencies is key. Core competencies age and rot, just like everything else. Destroying corporate culture. That's also critical. "Every business should be prepared to abandon everything it does." Those words from the generally mild-mannered Peter Drucker. And Mr. Drucker is right.

Design mindfulness is the core competence for tomorrow. That's what I've been selling. But that doesn't mean it doesn't have to be turned over, recharged, and revitalized. It doesn't mean that what it is that makes us special—from a design mindfulness standpoint—doesn't have to be redone, and redone regularly. Welcome the anarchist. Welcome the destroyer. We desperately need them.

33. Design mindfulness is more or less an enemy of TQM and re-engineering and customer-is-king. Our (American) newfound obsession with quality is of profound importance. Nonetheless, everybody's doing it. Great quality comes from most of Asia, from most of Latin America, and from most of everywhere. Planet-class quality is a necessity, but no more than a pass to the player's entrance to the stadium. It doesn't score points per se.

Moreover, there's a ton of evidence that the quality movement has, in effect, been kidnapped by the bureaucrats. Consider this description of the European quality standard, ISO 9000, from the director of corporate quality at Motorola:

> With ISO 9000 you can still have terrible processes and products. You can certify a manufacturer that makes life jackets from concrete, as long as those jackets are made according to the documented procedures and that the company provides the next of kin with instructions on how to complain about defects. That's absurd.

TQM can—and usually does—become mechanical. Most re-engineering activities are or become mechanical, top-down affairs. Customer-is-king can mean—usually does mean—slavish devotion to focus groups and market research, rear view mirror watching.

Maybe I can clarify my concerns, courtesy of two comments. The first comes from the respected health-care futurist Leland Kaiser. "When I meet a friend who has just returned from a visit to the hospital, clinic, or doctor's office," he wrote in the *Healthcare Forum Journal*, "I ask 'Did you have a good time?' This is the same question I might ask a friend if she or he just returned from a trip to Disneyland. A visit to a health-care facility should be a great experience."

Great experience. Amen!

Look, I'm in favor of "re-engineering the hospital," taking, say, twenty-four unnecessary steps out of a stupid, time-consuming, bureaucratic admissions check-in process. Doing that increases efficiency and customer satisfaction. But does it get, directly, to "Did you have a good time?" I'm not sure it does. In fact, I'm rather sure it doesn't. "Did you have a good time?" is an entirely different mind-set from slashing those twenty-four steps. One is about reduction (cut the unnecessary crap). One is about creation, holism. And that's where the great leaps take place in the market niche, in the marketplace, and in the world at large.

Another angle on it: the practice of the service of tea, called *chado* in Japan, can be a lifetime's occupation. Consider this excerpt from *The Book of Tea*, recounting a conversation between the ancient tea master Rikyu and his son, Sho-an:

> Rikyu was watching his son Sho-an as he swept and watered the garden path. "Not clean enough," said Rikyu when Sho-an had finished his task, and bade him try again. After a weary hour, the son turned to Rikyu: "Father, there is nothing more to be done. The steps have been washed for the third time, the stone planters and the trees are well sprinkled with water. Moss and lichens are shining with a fresh verdure; not a twig, not a leaf have I left on the ground."
>
> "Young fool," chided the tea master, "that is not the way a garden path should be swept." Saying this, Rikyu stepped into the garden, shook a

tree and scattered over the garden gold and crimson leaves, scraps of the brocade of autumn! What Rikyu demanded was not cleanliness alone, but the beautiful and the natural also.

Re-engineering, TQM, getting a bead on customer satisfaction; they're all about washing the steps for the third time. And the thirty-third time. But as one seminar participant said of his company's quality program, "We're shampooing the rug for the twenty-fourth time while the competition is busily going about pulling it out from under us."

TQM. Good stuff. Re-engineering. Good stuff. Customer-is-king. Good stuff. But they're not "Did you have a good time?" Not the shaking of the tree, the scattering over the garden the golden and crimson leaves, scraps of the brocade of autumn. That is, they are not design mindfulness.

34. Design mindfulness equals love. It may be a world of bits and bytes, of Nintendo offering our youngsters for $250, via its latest Gameboy, more computing power than the Strategic Air Command's boss had at his fingertips with the fastest Cray supercomputer just fifteen years ago. Still, design mindfulness is about being special. About the need for nothing more than your eyeballs to "know it when you see it." *All we need is love.* Design equals love. Okay?

35. Design mindfulness is living life out loud. The legendary Notre Dame English professor, Frank O'Malley, laid down some rules for teachers when dealing with students. At the top of the list: "Never accept a tepid response."

A game designer asked Nintendo's president, Hiroshi Yamauchi, "What do I make?"

Yamauchi's response: "Something great."

"If you ask me what I have come to do in this world," said Émile Zola, "I who am an artist, I will reply, 'I am here to live my life out loud.'"

Each of us here has been given a gift. We didn't ask for it. We don't deserve it. The gift? We are in positions of responsibility, positions of opportunity, at the time of the biggest change in the ways of commerce and business—indeed, of life itself—in the past several hundred years.

It's a madhouse out there. Everything is changing. Everything is being turned upside down. That is an extraordinary gift. An opportunity our parents, our grandparents, our great-grandparents, our great-great-grandparents didn't have. So the question is, are we up to the task? Are we willing to live our life as loudly as these very loud times ask for? If we're not, we're ignoring the gift, we're abdicating our responsibility to our peers and our customers and to, mostly, ourselves.

"I am here to live my life out loud." "No tepid response." "Something great." There's a message there, and a loud one, for all of us who are in pursuit of design mindfulness. Anybody listening?

II. The Redesign of Business

Design for Facilitation, Facilitation for Design: Managing Media to Manage Innovation
Michael Schrage

I was in Japan last week, and as I always do when I'm in Tokyo, I went to the AXIS Building, which is where they publish *AXIS* magazine and where they have many design studios and exhibits. So, I went to the AXIS Building. God knows I wasn't going to buy anything at eighty-five yen to the dollar; I just wanted to browse and window shop. And there was actually an exhibit there on design and business.

In the exhibit were some quotes from designers. I read them because they were in English, and I can't read *kanji*. Some of these quotes were particularly interesting, so I thought that I would start by letting the designers themselves describe the relationship between design and business.

The first one was from Raymond Lowey: "I am streamlining the field curve." I thought this was a great quote because it's so ambiguous. Are we making sales faster, or are we just making them skinnier?

The next one was from Dieter Rams: "Good design is as little design as possible." It's a sort of Teutonic Zen statement. It doesn't necessarily go over well when you're requesting an increase in your budget. Well, good design is less design, but my favorite quote that I came across was from one of the godfathers of American design, one of the great, wise, old men of American design, Buckminster Fuller: "You have to make up your mind to either make sense or make money if you want to be a designer."

And I feel that that's the not-so-false aspect of the design/business dichotomy. Lord knows we've been discussing it at the café and it's the connotation of much that I've been hearing here. The purpose of this talk is to try to come to grips with to what extent you can make sense and make money at the same time, without sacrificing too much sense or sacrificing too much money. Let's face it, the existing design/business model relationship doesn't work very well. The designers aren't happy.

46

The businesspeople aren't happy. We're not just talking about opportunities missed. We're talking about genuine disappointments, genuine failures, and the nature of the success and failure. If you'll excuse the language, the relationship between design and business really sucks. I think that goes a long way in determining the nature of this particular Aspen design conference.

I have more of a business background than a design background, although I'm fascinated by the process of creativity. The column that I write for the *Los Angeles Times* is called "Innovation." I'm fascinated by how people create change and by the tools and mechanisms and media that they use to create and manage change. I look at these things as much in the business context as in the societal context. I'd like to mention a number of things. This is standard stuff, so I'm going to get through it.

It isn't business that's the culprit these days. Business needs to do things faster, so things need to be done more quickly. For all the Southwest Airlines there are seeking to make things simpler, the trend in most businesses is towards a greater degree of complexity. And the challenge is often either how do we make complexity simple, or how do we develop systems to manage complexity and build bridges between inherently different parts of an organization with different vocabularies, different sensibilities—both design and financial?

The other important factor is the need to provide unique value. You know, if you don't have unique value at an appropriate site, you go elsewhere. And one of the great growth trends in design work, as we all know, is in out-sourcing of internal design departments. It's the same thing with computers, the MIS department trend is for out-sourcing unless you can demonstrate a unique value within the organization.

The other trend—and I'm prepared to say such a dirty word—is in politics. When I say politics— I don't mean Newt Gingrich and *The Contract for America*, I mean when times are not particularly good or we have fundamental differences in internal politics. I think it's fair to say, that most of us find greater degrees of internal politics in organizations rather than lesser degrees of internal politics.

So how do we deal with all of these things? I want to talk about my design bias in approaching the issue of redefining the relationship between business and design. Because we have to ask ourselves: How does design really create value in business? And the question that concerns everybody in terms of my design bias here, is how can designers productively redefine relationships with their clients, both internal and or external? Everything I want to talk about, and all the work that I've done, is pretty much harnessed to this kind of question.

What is my personal research bias in all of this? Several years ago, for reasons I will not bore you with, I wrote a book about collaboration. I was covering high technology and innovation as a newspaper reporter for the *Washington Post*. I began to realize, as I covered these Silicon Valley and Route 128 companies, that we have this model of great entrepreneurs, truly creative individuals coming up and overcoming insurmountable odds. And I discovered more and more that that kind of spirit was not the critical factor, because what I was really finding at the center of innovation was not just creative individuals, but creative relationships.

I'm very interested in creative relationships, collaborative relationships, so I have had the great good fortune to interview people like Watson and Crick. I did a lot of research into Braque and Picasso, Wilbur and Orville Wright, Hewlett and Packard, Rosnyack and Jobs, Mitch Kapor and John Sachs and the creation of Lotus 1-2-3, and Ray Ossie and his team in the development of Lotus Notes.

I was very interested in what made collaborative relationships different from other kinds of relationships. You look at a history of quantum physics and you find that there are towering geniuses by any measure. Yes these people are viewed as very effective collaborators. Advances in the field took place, not just through individual great minds generating new ideas, but through the interaction of individuals. The value was as much in the interaction as in the individual. And so I became very, very interested in the notion of collaboration. I wanted to study collaboration. I wanted to get an insight into how people create together, as opposed to our standard Cognitive Psychology model where to understand creativity we don't study creative relationships, we study cognition and see if we can imitate the thought processes of geniuses, when half of what we really want to imitate is the kind of relationships that those creative people have with their colleagues and their peers.

Now, what happened was that my research split into two parts, and this is the stuff that I'm working on at MIT. One is in the realm of prototyping. How do organizations build models of reality?

What are the median methods that organizations use to build models? How does Sony do it? How does 3M do it? What are the ingredients of putting together models of reality? How many times does one organization prototype something versus another? What insight does the way an organization prototypes give you into how that organization views creativity and innovation? What does what I call "the cultures of prototyping," say about that organization?

I've done a lot of work in the realm of prototyping, and I will continue to do work in that area. I have my standard shticks on that area, and I could give you my standard spiel; but, of course, I'm going to talk about the area I know less about, which is the one that I'm most curious about because my work in prototyping keeps driving me there: How do we manage the interactions that go on around those prototypes? How do we manage the changes in behavior that occur? How should we structure them? How should we band them?

These are the questions that are increasingly consuming my time and thought and energy; and I'm going to be presumptuous enough to say the more I learn about it the more confident I am in saying I think these kinds of questions should be consuming your time, your thoughts, and your energy also. One of the key things that I found in doing the work on collaboration is that the core at the center of effective collaboration is not so much the collaborative temperament or collaborative cognitive style, but the creation of a shared space.

A shared space can be when you're sitting across a table and you're scribbling stuff on a napkin or on a yellow pad and you're discussing an idea, and the person says, "Well, yeah," and they take the pad from you and they begin writing it down. All of a sudden you're discovering that

49

you're not so much talking to the other person directly; you're talking with them through the yellow pad, or you're talking through the models or the scientific experiments. You discover that the shared space really becomes the battleground or the creative ground for the kind of collaborations that you're trying to create, for the kind of creativity that you're trying to foster.

If you look at traditional communications theory, the standard model is S/R, and I'm sorry to say the bulk of the meetings and conversations that we have fit the classic model. For you Pavlovians in the audience, S/R does not stand for Stimulus/Response. It stands for Senders/ Receivers. That's pretty much the standard model of communication if you look at Information Theory, Shannon-Weaver, or Cybernetics. In all three, Senders/Receivers is basically the model of an exchange of information, messages, bits, and bytes.

Where creativity and collaboration truly occur is in conversation that occurs, that is mediated and managed through shared space, whatever that shared space may be.

One of the most striking experiences I had was when I was a reporter at the *Washington Post*. The *Washington Post*, when I was there, was such a bloody cheap organization. We had a text editing system, and they didn't want to buy keyboards for every reporter because you weren't writing all the time. Of course they didn't understand the concept of peak time, because sometime between the hour of 4:30 and 6:00 P.M. we were all writing. The owners would come in at eleven in the morning and be surprised that nobody was using the keyboards, so they thought we had an excess. Go figure. When I was the technology person and we were covering semiconductor trade issues. I had the background in semiconductors, somebody else was covering trade, so we collaborated on the story.

The typical way we would do it is, I'd use the keyboard, write my story, send it off, and have him write. It would be the Send/Receive model. We would exchange messages. So one time the news broke just before deadline. When I say deadline I mean production deadline—you don't get it in you're beaten by the *New York Times* and the *Wall Street*

Journal. That can't happen. What happened was, there was a lot of news that day, so both of us had to literally sit side by side at the keyboard. (The *Post* did by the way expend the extra money, they put the keyboards on wheels so you could actually shift them back and forth.) And what happened was that we began talking about the story.

I began writing and then Steve took the thing and he began writing, and it was like, "your turn to drive; my turn to drive." It was very interesting because—and I remember this very explicitly—you know how it is when you have an out-of-body experience, and you watch yourself doing something? I remember the feeling that I was not talking with him anymore as much as I was talking with him or communicating with him through the screen. Our conversation was driving what was on the shared space, and what was on the screen was driving the conversation about the story we were writing. That was a fundamental shift. I gather it doesn't quite seem a leap of the imagination, but this is when I began to become interested, not just in the notion of collaboration, but in the tools, the methods that support collaboration, and the nature of interaction and shared space.

If you want to talk about one of the most interesting shared spaces in history, you can talk about Watson and Crick and the double helix. If any of you have read *Double Helix*, it's a terrific book, both about creativity and collaboration and shared space and prototyping. They were working with organic chemists, x-ray crystallographers.

Do you know how many experiments Watson and Crick did on their way to discovering the double helix? They did absolutely no experiments. Not one. All they did was take other people's data, synthesize it, sketch helixes on blackboards and then build metal models of helixes. That was the big deal, building metal models, because that way they could test their hypothesis of whether the bases matched and what have you. So here is Jim Watson, out of Indiana University at the age of twenty-five; he wins the Nobel Prize. He's the co-winner of the Nobel Prize for discovering DNA. He didn't get his fingernails dirty once in lab experiments. So managing shared space effectively can clearly be important in creativity and innovation. It's the relationship between Watson and Crick and Louise Wilkins that made the huge difference.

The key thing that I came away with is the notion that it takes shared space to create shared understandings. When you talk about people speaking different languages, be they foreign languages or just different subcultures of business and design, it's tough to create shared understandings with that standard Sender/Receiver model. It really is tough. Where is the shared space?

Whatever property is involved is the shared space. And that's what I became convinced of in prototypes. Because if you change the properties of the prototypes, even prototypes in cardboard, you're changing the quality of the collaboration. Industrial designers may think I'm simply rearticulating something you already know but never may have bothered to explain to people. Doing a prototype on a calculator or anything in Renshape foam is a different process than doing that prototype in cardboard. Doing a CAD prototype is different than building a clay model. What's the dialogue between those prototypes?

I became very interested again in the question of how do we manage prototype or shared space? What's the new kind of prototype when somebody comes up with an idea? How many cycles of prototypes do we go through? How many different types of media are we prototyping with? What's the difference between the way GM prototypes a new product versus the way Sony does it? What's the difference between the ways Sony and Matsushita do it? What can you say about Ford versus Toyota from the way that they use clay models in creating a dialogue between their cab infrastructure and their solid-model CAD infrastructure.

I thought these were very interesting questions, because they really gave me insight.

I will now say something that I'm not particularly proud of. One of the reasons why I was personally interested in prototyping of the research medium is that you could look at the Δ—the changes in prototypes— without talking to anybody, because those prototypes spoke for themselves. And that's what I love so much about electronic media, which I'll come back to in a moment. You get to see different versions and how the versions are, and you don't have to interview people. You can see the changes. You can create objective measures, which at places like MIT, more so than Harvard, as John well knows, are hard numbers.

But there's something else you discover when you look at this. When I look at these things, the classic model of creativity and innovation is: innovative teams create a generation of innovative prototypes. That's sort of the model. Perfectly logical, that's why people are concerned with self-managed teams. What's the right team to create this product? But you know what I've discovered? When you talk about teams that are really, really successful and have a wonderful track record in these areas, you find—when you really talk with people and when you really observe what it is that they're doing—that the prototype is somehow influencing of team.

People see something, they like it, they may not necessarily be in the same department or same group and they say, "Hmm, that's interesting. Hadn't thought about doing it this way," or "What if you did it this way." So all of a sudden, instead of the team driving the development of the prototype, you have the prototype driving the innovation process. It becomes the medium of community.

It's the same kind of thing John Kao was talking about earlier in relation to what he wanted to do with Lotus Notes. People see things; they get some sort of sense that, "Hey, this is interesting, there's a way that I may be able to add value to it," and it becomes part of the community property.

By the way I just want to cycle back and talk about shared space. Too often in design and in science, people don't want to share. That's not to say there isn't a value to proprietary space, but for fostering certain kinds of creative collaborations, you need the shared space. And in order to have innovative prototypes generate innovative schemes, you need those prototypes to be shared spaces. So what I discovered despite my best efforts to keep the human factor out of my work in prototypes, was that it was really, really interesting the way that shared space worked.

I love looking at the media of prototypes, the different media, particularly virtual media in a shared space. But there was an ecology of events going on here, and that ecology of events was that new kinds of media created new kinds of interaction. I had to ask questions like: Do we just let that interaction happen? Are there ways of leveraging that kind of interaction to create new kinds of conversations and new kinds of value? And that really begged the question: How do you design the productive interactions? And this is the kind of direction that I want to see pursued. I don't have the answer. I have some approaches.

I used to be focused on how do we create really interesting opportunities for prototyping, which is a damn good question, and we have damn interesting results. Unfortunately, to get a really good answer to those kinds of questions, you've got to ask this question: How can we design really productive interactions? What do we really mean when we talk about productive interactions? It doesn't mean, for god's sake, let's have more meetings. One of the most interesting areas that one can dwell upon is the realm of facilitation.

I've been in meetings that have been very well facilitated and meeting that were very poorly facilitated. In the well-facilitated ones people walked out feeling, "We need to have this kind of experience more often." But, of course, it was the exception rather than the rule because hiring external facilitation is often very expensive. But if you look at books like *How to Make Meetings Work* by Dayle Straus, you'll find that there's a very interesting body of literature on facilitation.

That body of work has a great deal to say about facilitation in relation to prototypes and models. Some of it is: How do we tap the collective ideas of people? But the idea of facilitation around objects in relation to certain kinds of "stuff," they don't do that kind of work. They've got the studies of meetings being conducted independent of technological support. These are just new areas. But at the same time I felt like the work on facilitation could inform the work on prototyping and building models. Because what the facilitators do is create interaction. Good facilitators create good interactions. Again, I think that's increasingly going to be the design question that designers will face. Not just what kind of products and services do we want to design better, but how do we create interactions that boost the odds of flushing out good ideas, good approaches, good questions?

I'll tell you what I think should and shouldn't be. I don't believe in the kind of facilitation that says, "Tell me how you feel." I've unfortunately witnessed too many facilitated meetings that become gripe sessions. I'm talking about productive interactions, not $95-an-hour, are you a Freudian or a Jungian? I'm talking about something else, which is, what kind of results do we want from this meeting? This is the kind of question that people have to ask themselves. *The goal of facilitation is to design and create interactions that add value.*

I think that we really need to be honest and ask ourselves: Is this how we run our creative processes and our innovation processes in our organizations or with our clients? Because—and by the way, I don't think there's anything wrong with this—most people are concerned with: How do we solve the problem most effectively? What's the beef? How do we solve the problem? Oftentimes the kind of interactions that we need in order to create are secondarily or tertiarily related to the way that we design our products because we're so focused on solving the problems or creating that opportunity. To what extent are we allowing our prototypes and our models to drive interactions by default rather than by design? To what extent does facilitation and management, managing and structuring interactions, give us unique opportunities to add value?

I haven't done formal research on this and when I don't do formal research on something I do informal research. There are a bunch of companies I have a fairly good relationship with, and I called and asked them a basically simple question. All the companies are fairly high-tech, very customer-sensitive; they have a very good focus group, and they really care about soliciting customer-feedback. I asked them this question, "Have you all ever done an internal focus group about how you felt the product you're designing was going, facilitated by the same kind of person you hired to handle consumer research, and anything like that?" And none of them had.

By the way, these are all companies that are concerned about improving the quality of their relationships in cross-functional development of the products and services that they do. I think it'd be an interesting experiment to go to your marketing research department or marketing department and say, "Let's run an internal focus group just to review the status of a product that we're developing. What's being done well? Where are we satisfied? Where are we not? Let's use this kind of facilitative mechanism to be introspective." I'm very curious, and I hope that some of you would be very curious, to find out whether that focus group would yield results that reinforce certain injuries, or yield interesting insights into ways you could add value to what you're doing or add value to the interactions involved in the design process.

I don't think organizations are doing this sort of thing. They're so busy and intent on what Tom Peters was saying the other night about how they're perhaps listening too much to the customers. We're not taking advantage of the same techniques we use to design better products for our customers, to take better advantage of our own resources and capabilities. That doesn't strike me as an illogical or an unfair assertion. So what are the ends of facilitation in that environment? Well, there is brainstorming, generating ideas. At this point I have to add a story.

One gentlemen that I just spoke with before the talk said that in some of the meetings he's facilitated, when you're in a brainstorming session you're not supposed to criticize ideas, you're just supposed to let them flow. And this gentlemen said that one of the ways, one of the little gimmicks—and I like this kind of gimmick—is that you give everybody a Nerf Ball when they come in and if somebody says anything critical you throw the Nerf Ball at them. You know, it's a way of getting people involved. They actually pay attention to whatever people say so that they can line up for a Nerf Ball session.

My best friend, who runs a multimedia company in Silicon Valley, says his firm has a different approach to brainstorming. They have water pistols so that you can squirt people. The only problem was—and forgive me for saying this—that there was a rumor going around that some people don't use water. So they're going to have to change the way they handle that kind of problem.

There have been some interesting academic studies that have shown that facilitation does make an enormous difference in both the quantity and quality of ideas in professions. To what extent are you soliciting your ideas? To what extent do you want to generate, to evoke ideas or just provoke different kinds of conversations that haven't existed before in the organization? To what extent are you using facilitation to achieve consensus rather than to identify genuine conflicts? Everyone knows people have "meetings," and, of course, 90 percent of the meetings have no agenda, but the worst is that they lump in three or four of these things simultaneously. It's absolutely bizarre. And people are always surprised that it's a waste of time.

Just one other aside here. I deal with people all the time who talk about how videoconferencing is going to save time. It's going to save time on travel. It will be more effective and efficient. Every single person I've talked with, every white-collar worker I've asked, "What's the biggest waste of your time?" has said, "Meetings." And then they talk about videoconferencing. Geez, why would we recreate in cyberspace the single biggest waste of time we have in the physical world? It's absolutely bizarre. But again the issue is: How explicit are we being on these things? Are we cheating ourselves because we are taking it for granted that because we all know each other something productive will result?

There's actually a book, *Conflict of Consensus: A General Theory of Collective Decisions* by Mokovichi and Dois, reviewing the social-psychology literature on consensus and conflict and the research on how groups make decisions. And the fact is, there are genuine techniques and genuine approaches that make a difference in identifying these things, if we have the brains to design these interactions rather than default them. Then, of course, there are facilitative aims where you try to encourage participation and seek buy-in. Sometimes you facilitate for political reasons. Good for you.

What are the various techniques? Well, we've talked about squirt guns and Nerf Balls, but there's also the number of people and the mix of people that you have. I'm amazed that facilitators will stuff thirty people into a meeting. Why not just have five? How about if you decide to have meetings of ten? What if you decide to sequence things differently? How are people determining the number and mix of people? Is it just done in a certain way because that's the way we've always done it? We've always had a representative from this department. We've always had this number. It's not officially a meeting. Well then, for god's sake don't run it as a meeting. Run it as something else.

Then there's the issue of time. I think one of the best ways to run a facilitation is to set the time and stick to it. It's sort of bizarre. This is one of the things that perhaps people know they should do or don't really know they should do but some organizations are increasingly doing it. How are we changing the environment to structure interactions? You want to keep people away? You keep the thermostat down. The best way to terminate a meeting? Make sure the thermostat is broken and the temperature's too high. Best way to cut off a meeting. No question. You want a quick meeting? No chairs. How many of you have been in a technical-review, no-chairs meeting? See, I've got under ten people doing this.

Now, with all due respect, there needs to be a little more experimentation in this area. Even if you discover it's not the way you want to run a meeting, that's a useful data point to have. Have that kind of mix in structuring an interaction. There is a difference between a meeting at a round table and a meeting at long, oblong tables, or using white boards and not using white boards, or having computer-augmented meetings or not. There's the use of role-playing, getting people to play certain roles. How many of you here do that kind of role-playing where you play the role of either a supplier or a customer in the discussion of a technical review of a product or prototype you create? Role-playing? Gee, maybe five or six more people.

It kind of concerns me that people here are much more concerned with playing with their prototypes and their products than the way they interact in relation to the design environment. This is the crème de la crème of the design community. I think there needs to be a bit more experimentation in the interactions you structure among yourselves and the people to whom you purport to add value. Now, of course, there's the issue of facilitation for internal versus external. How do we do this? Do we have honest brokers within the system? Do you know what you call a really good internal facilitator? A good manager. They tend not to be objective.

My favorite facilitation issue is one that I came across in the groupware, which is the issue of anonymity versus attribution. I was working with an aerospace company with a lot of serious design issues. This is the classic example. You know the software phrase, "they turn the bug into a feature"? What they did was, at a brainstorming session they linked up twenty or so IBM computers to a shared screen. You could type and all this stuff would be shuttled into the screen, and because the software wasn't sophisticated enough to identify who made the contribution, they could say, "Well, we can have anonymous meetings now." And this was one of the *virtues* that sold the software to this large aerospace company.

I was doing some work with them on this issue of groupware technologies to support interactions. And I got into this argument with them and said, "Look, people should attach their names to their criticism and comments." And this person patted me on the head and said, "No, no, Michael, you don't understand. If you did that sort of thing you'd lose your job. You'd alienate people."

And my argument was that they were using technology to subsidize a value in the organization that they might really want to change. And that perhaps in the longer run it would be *healthier* to have a culture in which people could attach their names to their criticisms rather than rely on the mask of anonymity to make their contributions. And this is a question that you all are going to face if and when you follow up on these issues of facilitation. If you are in an organization where the best suggestions for improving products and services and approaching innovation are made anonymously rather than by people who are proud to attach their names to their comments and criticisms, you've got a pathologically sick organization. But you know what? If you've got a pathologically sick organization, you've got the technologies that can let you paper it over. You can turn a bug into a feature.

Then again, if you're structuring interactions—here's the great thing about the technology and the bad thing about the technology—the more choices you have the more your values matter. Ten years ago the technologies didn't give you much choice, it didn't matter what your values were. Your value was you bought the technology or you didn't. Now with a blank piece of paper, you can write on it, you can put a photo on it, you can sketch on it, or you can fold it like an origami bird. You have those same kinds of issues springing up into the design of using technologies to structure facilitations and interactions. Now that's finished.

Let's go back to what I was talking about in the business context. What are the concerns of business? Speed. A good facilitator, good facilitation, is a way of accelerating the process of creating shared understandings. Facilitation can be an important tool for accelerating the product or process of innovation.

You want to cope with complexity? Facilitation is a way of getting people in the room to create a shared space to create their shared understandings, for value creation. Now, here's what I think a lot of designers would like. I think it would be very interesting if people in the organization felt, "Geez, we could have this new product meeting—we could have this technical review, but, you know, if the people from design aren't there to facilitate and manage the prototype and the interactions, those interactions aren't going to be as good. We will have a qualitative loss in how we talk about it, how we think about it and how we do it if they're not there." It's a unique way of creating value. And that's the switching costs for an organization.

Right now there's clearly a market in efficiency in regard to coordinating these things. Who's handling this facilitative task? No one. No one is officially designated to do this. You know, there is a shared space that we all know the managers and the general managers oversee; so they legitimately need shared space, they don't need to market. But the shared space that matters in most organizations is in the technology of the budget and the spreadsheet. That's the shared space that matters. If you're in companies that create products and services, you damned well better come up with shared spaces to have parity or at least have an argument to rival the importance of the spreadsheet as a technology in shared space.

And finally there's the issue of politics. Well, you know, I think that one of the reasons—and it's my personal opinion, not research—why designers are isolated is their solution to solving these things. If you're facilitating a meeting effectively and you manage to get a buy-in on various things, if it's a failure then it's everybody's failure. There's a way of managing the politics of an organization. It's a way of insinuating yourself into how the organization handles its political capital, not just its economic and innovation capital. Facilitation can be a powerful political tool as well as all these good things. In fact, when push comes to shove, I'm prepared to be cynical on these things but well-managed facilitation is a good tool for cynics as well.

I just want to briefly touch upon the future of facilitation. Increasingly, as we know from Lotus Notes, computer-aided design, and computer-aided engineering, the trend is for more and more virtual prototyping, more and more distributive development of things. And some of the work that I'm doing is concerned with how to facilitate a collaboration on a network. Do we use software to automate the facilitation process? Or do we give facilitators software tools to augment the facilitation process? Do we facilitate in real time? How do we do asynchronous facilitation?

These are some of the issues that Hampden-Turner may address with his mind maps on a piece of paper. What happens when you do mind maps in virtual reality? Will we use software agents to manage facilitations? Will we be able to layer through and see who made what changes on a prototype when using these things? And will we be able to change the conversations we have in relation to that virtual prototype, as a result of these kinds of agents?

So there's not just a need for structured interaction and facilitation in the physical world. I think the technological trends are going to push us to examine the issue of structuring interactions in the virtual world as well. In other words, these are questions we can't avoid no matter which way we turn. And that's why I care so much about them.

Now the obvious question after this talk on facilitation is: Why designers? Why should designers do this? And of course the obvious answer is: Why the heck not? Realistically, who else? Who else in the organization? Why? Because this is what I think is the value proposition for design over time, the creation and management of shared spaces and the interaction that surrounds them. Because this is what you guys do anyway.

You know the cheating, sleazy thing is that what I've done in my work in prototyping is I've looked at what people have done and I've just formalized it. The best designers, the most effective ones, do these kinds of things anyway. They're very good at creating and managing shared space. Who has the technical expertise to create shared space? Who is used to going around various parts of the organization and interacting and learning what the trade-offs are? Be it graphic design or industrial design, certainly in software development, it's the designers.

I think designers are uniquely positioned to become the stewards of shared space, be it physical shared space or virtual shared space, and that is a way of uniquely creating value in the organization. Because how organizations create shared space is how they create value. That's what I learned from the work that I've done in collaboration. That's my bias. And you're welcome to disagree with it but, boy, there's an awful lot of empirical evidence that I think lends weight to the perspective I'm trying to present here today.

So here's some of the questions that I think are important. When you're with your clients, be they internal or external, if you care about collaboration, if you care about joint creativity, the first question you've got to ask after "What's the problem?" is "Where's the shared space?" Because chances are if there ain't shared space, you ain't collaborating. You may be having a great conversation and you may have fun shooting the breeze, you may come up with great ideas, but where's the creativity occurring in the shared space?

The second question, the one that I'm going to be spending a lot more time researching, is what kinds of interactions and conversations are we trying to create around the shared space? When do we want the interactions and the conversations to drive the prototype, the model development? When do we want the prototype and the models to drive the interactions and the conversations? That's a strategic question, not just a tactical question for innovation management. Are we creating and properly managing the shared spaces that matter in our organizations? If the answer to that question becomes yes, more and more, designers are going to be more than happy with the amount of power and influence that they have in the organization. And the people in the business part of the organization are going to inherently recognize it. Why? Because it's a shared space.

Now there is a downside to this, because I cannot look you all in the eye and say, "Does this kind of approach lead to better design?" I can't tell you yes. I don't know. My view is I think it loads the dice. I think it creates the context that really changes how the organization perceives of itself as critical. I think it's the sort of thing that we need to energetically and aggressively and creatively explore.

When I was chatting this out, somebody who is cynical was suggesting that to some extent all I'm doing is calling for a repackaging of stuff that's already done. I would like to be less cynical. What I'm asking people here to do is to rediscover the principles that gave design, that gave creative people, their influence in the first place, which is at the core of these shared spaces, these creative interactions, that people wanted to be a part of. And I think that that's what I'm really asking you all to do. To rediscover the things that you do well and begin to be introspective in ways that enable you to take advantage of the core competencies they really bring to these kinds of opportunities.

I think it's a tremendous opportunity. And I think that if you really look at facilitation as a medium to add value to shared space that I think it's going to have an enormous impact on the business relationship between the people who design the value and the people who pay the bills.

Managing the Virtual Company
Hatim Tyabji

I will be talking this morning as a practitioner, somebody who is an operating man. Nothing that I'll be talking about is something that we are *thinking* of doing. Everything I'll be talking about is what we *are* doing. A lot of what we're doing is very relevant to the design community, and so I'll try to contextualize it from that standpoint.

One of the key things we tried to do when we started the company, VeriFone, was to create an environment in which to foster creativity and innovation. That was, and is, the fundamental basis for designing the operating structure of the company. John Kao says, "The Zen Buddhists call for the purity of a 'beginner's mind' as a path to genuine creativity." Very often I am told that if you started the company you had the "beginner's mind." But the fact of the matter is that as the company progresses it no longer remains a "beginner's mind." The question becomes, how do you retain that purity so you have an atmosphere of continuous creativity? To encourage this atmosphere, we put together five goals. I'm going to read them to you, and then I'm going to indicate how we have made those five goals an article of faith in our company— and why.

The five goals, or imperatives, were put together right at the company's beginning in the mid-eighties. The first was that we were going to form a company that would sharply improve quality-of-life for our people. That was a very, very key goal for us.

The second key goal was that the company would go to where the talent is. Generally, if you want to join a company, you say, "Well, I've got to move to Austin, Texas," or San Francisco or Paris or wherever the company happens to be based. Instead, we go to where the talent is. I'll talk more about that, about our extreme decentralization and how we have been successful doing that.

The third key imperative was to effect a quantum increase in time-to-market in our product development, and to maximize qualitative design attributes. It is one thing to go fast to market. However, it is no fun if your product is not going to be exciting. So the two had to go in tandem.

The fourth imperative was to compete globally in an ever-shrinking world. We decided this in 1986, at a time when the company was operating only in the western part of the United States. We had no operations east of the Mississippi, and most of us looked upon New York very much as a foreign country. To some extent most of us still do.

The fifth imperative was to form a company that would have insensitivity to time, to distance, and to location. In other words, it shouldn't really make any difference what the time zones were, what distances one was traveling, or where one was based. The important thing was to have the right people in place.

So we put these five imperatives together and the question was: How are we going to bring this to life? We did this in a couple of different ways. We put together a program that we call "Commitment to Excellence," which consists of three fundamental parts. First is a strong philosophical framework that we call the "VeriFone Philosophy." Second are the strategic elements that underlie the philosophy or that give flesh, so to speak, to the philosophy, which we call "Excellence in Thought." And the last item is "Excellence in Action." Because, you know, all the philosophies and the strategies in the world don't make any difference if you don't execute them and bring them to life.

So in this context, we put the philosophy together in a book. There are eight fundamental precepts in the philosophy. In the interest of time, I'm not going to read those precepts, but they are precepts that we live by. These are not something that we just put on a nice piece of paper; these are very much the heartbeat of the company. Each of these eight precepts is fleshed out in a separate page.

You might say this booklet is a little thick for eight precepts. Well, that's because this book is in eight languages. We are a global company. We are totally decentralized. It is not possible to ask your people to follow you if you don't talk to them in their own language. One of the strongest elements of our global orientation is that we respect people's differences. Not only do we respect people's differences, we take advantage of people's differences. This philosophy has become a very strong article of faith.

The second major item is what I call "Excellence in Thought." "Excellence in Thought" is really a continuum of strategies. These are a collection of electronic mail messages that we have written over the years. To share with you some of our thinking and the way we do things, I have excerpted three subjects to read from those e-mails—just brief extracts to give you a sense of the kind of people we are and the kind of company we are.

The first one I'm going to read, from our "Excellence in Thought" series, is one I happened to write. The subject is leadership, written in June of 1991. This e-mail was sent to I_STAFF, which means that it went to every man, woman, and child in the company, wherever that person happens to be based, instantaneously. I'll just read one paragraph.

> As many of you know, I have high standards of leadership. I practice these standards myself. I judge my own performance and the performance of others by them. I am not shy about advising others when they fail to meet my standards. By the same token, I fully expect others to advise me when I fail to live up to my own beliefs.

And then I've listed what the beliefs are. This is an example of what we do from a leadership standpoint.

The other is an e-mail that was written in September of 1991 on the subject of organizational effectiveness. All of you, I expect, either work in companies, run your own companies, or have at some point in time been part of companies. As you know, there are always changes in organization. I think any one of us, whether you agree with the change in organization or whether you disagree, would be far more accepting of the change if you understood the rationale for it. It becomes very frustrating when these changes come from on high, and you don't understand what in Sam Hill is going on. If you understand what is going on, you may disagree with it but say, "Fine, we have a difference of opinion."

So I wrote an e-mail explaining the philosophy behind any organizational changes we make. This was sent to every person in the company, and the message conveyed was that when an organizational change is to be made, there is an operative rationale for that change. And if the change doesn't fit within the fundamental framework laid out in this e-mail, then talk to me. Tell me that there are strong reasons for your beliefs, and we will change. From that context, I talk about the fact that today, in order to achieve organizational effectiveness, we must be able to take complex matters and distill from them a set of simple principles. This requires clarity of thought in formulation, consistency and frequency in communication, and an intensity of focus in implementation.

While each one of these is very important, the consistency and frequency of communication is really crucial. This is one of the things we all know, particularly from a design standpoint, but also really in any discipline. People generally don't hear you the first time you talk to them. The fifteenth time they may perhaps listen to you. It takes a very long time because we're all human beings. And one of the things that we have done in trying to design organizations, is to accept the fact that human beings are human beings. You are not going to be able to change human nature, you're going to have to adapt to human nature.

So, we've articulated five key principles for structural organizational effectiveness: (1) moving the point of power; (2) thinking globally and acting locally; (3) focus and accountability; (4) questioning paradigms; (5) strategic consistency and operational flexibility.

I go on to flesh out each one of these in specific detail, so people can understand that these are not just fancy words; there's a lot of thinking behind them.

I'm going to talk about just two principles. The first is "think globally, act locally." This is something that I feel very, very strongly about, so I will just read the e-mail. It says,

The attempted coup in the USSR was covered live on US television providing the viewer in Greensville, Ohio with more information about what was happening than Mikhail Gorbachev had in his vacation villa in the Crimea.

Which is absolutely true. You might, also by the way, have this burning desire to know why I picked on Greensville, Ohio. The answer is, I don't have the foggiest idea. It just sounded right so I put it there.

The immediacy and suddenness of such events reminds us thoughtfully of the world's shrinkage and our global citizenship. An organization can ignore fundamental changes only at its peril. In the coming decades we will see the evolution of a totally interconnected village. As an organization we will proactively locate ourselves throughout the globe without regard to geographic or political boundaries. Moreover, we will leverage our global advantage to position ourselves strongly for continued growth. As you know, this is by no means a new concept at VeriFone.

So what is the global concept at VeriFone? We talk about creativity. The fact of the matter is that the way we engender creativity is to have development and design centers all over the world. Each design center is designated as a center for excellence, and they do design, not for any particular part of the world, but for the entire world. Since the entire company, 100 percent, is totally networked, and since the entire company, 100 percent—no exceptions—works on-line, we are able pick each other's brains and to do it in such a manner that it doesn't really make any difference where any person is based. More to the point, we don't care where they are based. This is the really important thing, which has really fostered an atmosphere of intense creativity.

As an example, we have centers in northern California that are responsible for our systems development. We have a center in Honolulu, Hawaii, which is also responsible for some other major elements of systems. We have a center in Taipei which is the center of excellence for all electro-mechanical devices. We have a center in Bangalore, India, which has the responsibility for networking and communications software. We have a center in Paris, France, which has the responsibility for intelligent cards and security. Each one of these centers is totally connected. And the designers in every one of these centers are working in an extremely cohesive manner. This is not something that we are talking about doing; this is something that we have been doing for a while. If anything, we are continuing to build on this.

Creativity is something that one worries about all the time. And I guess in the context of one of those times when I was particularly worried, I used a metaphor that became very controversial in the company. Some people didn't like it. I put it in the e-mail anyway. The subject of the e-mail was "A Vital Message." It started off by saying,

> All of you have experienced, firsthand, the passionate intensity of my commitment to VeriFone's success. Recently, I have been troubled by certain basic issues of faith that seem to be clouding our future. As always, I am sharing my concerns with those VeriFoners—which is every person in the company—who can and who will resolve them.

And then I gave my metaphor.

> Companies are like tribal societies, composed of warriors and farmers. Warriors perpetually conquer new land to enable the tribe to continue growing. Farmers work the existing land to feed the tribe. At VeriFone, the warriors pursue new products, new customers, new markets, new distribution channels. Make no mistake, VeriFone is a warrior company. The warriors will drive our continued growth. Each of us must decide which role we are committed to play. The requirement for continued success is clear. This is not a spectator sport. The warrior does not respond to the competition; the warrior forces the competition to respond.

And that, fundamentally, has been the background of VeriFone. We did not join an existing industry. The words *transaction automation,* and the industry they represent, did not exist until the company was founded. We have created the industry; and, so far, we have been successful in leading it. However, we are very mindful of the fact that success can be very transitory, and we have always believed that the more successful we are the harder we're going to run.

These are the fundamental training books. Now the question is, is there any substance behind these training books? Are we actually doing these things? How do we work in a design context? I spoke about our global corporate citizenship. This is something very new. Generally speaking, you look at a company and you say, "This company is an American company," or "This company is a German company," or a Japanese company, or a Dutch company, or what have you. We take exception to that from a global standpoint, and we characterize VeriFone as a global company that happens to be registered in Delaware.

And though you might say, "Well, that's semantics," the answer is, it's not really semantics. Because when you think of yourself as a global citizen, and not as an American company or a French company, it is amazing how much the mind-set toward the company changes. This is a fundamental analysis; technology is an enabling function. Technology is important, and, by God, we use the technology. But if the attitude is not there and the mind-set is not there, I don't care what the technology is, it will become a disaster. And I don't have time for disasters.

To give you an example of how this mind-set translates into action, let me share this vignette. We had a good year in 1994, and so I got together with my staff in January and said, "Well, what are we going to do in terms of bonuses?" And the staff said, and I was very proud of them by the way, "We don't want you to do anything for us unless we have taken care of all the people in our company. We haven't given a company-wide bonus for a very long period of time. Let's do it this year."

I said, "This is fine. This is the bonus pool. Come back to me with your recommendations." They came back with recommendations. Given the amount of money we had for cash bonuses, we would not be able to afford to give any meaningful amount to the senior staff and have any meaningful amount left over for our people. So we decided that the entire senior staff was going to forego all cash bonuses, and we were going to take all of the cash and spread it out across all of the people in the company, obviously on a worldwide basis.

That was one element of the bonus, but the second element of the bonus, to my mind, was even more interesting. From the outset we have talked about ourselves as the VeriFone family, and at face value that may sound corny. But, do people really feel that way? Do people really behave that way? The answer is yes. When I talked about the second part of the bonus, we decided that we were going to close the entire company down, every man, woman and child, worldwide, for two full days. The reason we were going to do that was as follows. I'll just read this one short paragraph from the e-mail when I announced that.

> Take the two days off in April to spend time with your family. Your loved ones have supported you through 1994 and therefore supported VeriFone's performance. By all means the celebration of this performance should be shared with them because we look upon our families as an extended part of the VeriFone family.

Everybody was excited and so we asked, "Which two days are we going to take?" We decided to take the Thursday and Friday before Easter Sunday because it connected the holidays and people would have four days off. That sounds great, but most of the countries in Europe are closed on that Thursday and Friday anyway. And oh, by the way, Singapore happens to be closed on Friday also. So I said that if we announced that the company is closed on Thursday and Friday, those folks who are not based in the United States are going to say, "Whoop-dee-do, we were going to be off anyway, so what the hell are you telling us, Tyabji?"

So we announced that in the United States the two days, Thursday and Friday, would be off. For all countries outside the United States, local management would decide which two days contiguous with this Thursday and Friday would be off. So as an example, Spain—which is off more than it is on for those of you who have not worked there—was closed Tuesday, Wednesday, Thursday, Friday, Saturday, and Sunday. So for the bonus, Spain took the following Monday and Tuesday off. And I gave up all hope of any productivity for the quarter. The U.K. was off on, I believe, Good Friday and Easter Monday. So the U.K. took Thursday off and Tuesday off.

The point of mentioning this to you is not that we did anything very unique or very spectacular. It is to point out to you that we have a mind-set, and that mind-set is so deeply ingrained in the company, that whatever decisions we make are tempered by it. That's really the only point I wanted to make.

I'll give you two more vignettes. These are from a design standpoint in the context of this mind-set.

Many of you will recall that about two-and-a-half years ago the Miami area was hit by a devastating storm called Hurricane Andrew. We have a lot of people in the Miami area, and we were very concerned about what was going to happen to our people during that storm. And I was very, very proud that when the storm subsided, a lot of people in the VeriFone family, totally unsolicited, sent stuff, supportive e-mails, and went down there and helped them out. There was a lady in one of our groups, in Honolulu as it turns out, who took it upon herself *personally* to go there and make sure that the members of the VeriFone family were taken care of. This is an impossible job, terribly emotional, if I say so myself.

So when we were going to design an annual report for that year, I sat down with our designers. By the way, design, to my mind, is so broad and so important that I am personally involved. So I asked the designers, "How are we going to design the 1992 annual report? What should we do?" And one of our designers said, "Hatim, we know how strongly you feel about people. We know how strongly you feel about the VeriFone family. Let's dedicate this annual report to our people." I said fine. Talk to me. What does that mean? How are we going to do this? And they made out the outline and we said, "Let's do it."

Our annual report that year was dedicated to VeriFoners. In order to do that we put this particular lady who did that super job in Miami on the cover along with some other people; and there was a front page letter talking about what she did, how she took care of our people, and how good we felt about her. And then we sent two photographers with their cameras around the world to our VeriFone locations. They didn't take shots in a vacuum. They they talked to the managers, they found out who the real people were who were making things happen. They came back with vignettes and they came back with photographs. Let me tell you something, that annual report was superb. It was superb because it reflected the context of people.

Another vignette shows how we use design and industrial design to tremendous competitive advantage. Many of our products are used largely in a business-to-business environment. About a year-and-a-half ago we decided we were going to take a major plunge and go into consumer marketing. Well, what do you do when you go into that kind of a situation? The ergonomics of the products change dramatically. The touch and feel and user interfaces obviously become very, very crucial. We were mindful of that. We knew we didn't have all the answers.

So I sat down with the design people and said, "How are you going to do this right? If this thing doesn't look right, doesn't feel right, it won't go anywhere. And none of us is going to be very happy. The software and all that may be fine, but the touch and the feel are so crucial. How are we going to ascertain that?"

We talked about a variety of things, and we certainly talked about doing focus groups and some of the classical methods. But in addition to that, one of the designers came back and said, "Hatim, I've got a crazy idea." I said, "OK." He said, "Let's buy a Winnebago." I said, "You're cuckoo." He said, "No, let's buy a Winnebago. And three designers will travel for six months in the Winnebago. They're going to go to small towns, and they're going to go to big cities, and they're going to go to shopping malls and other areas where ordinary people go in and out, buying groceries, buying this, buying that in the context of their day-to-day commerce."

I said, "We'll have the Winnebago and we will call it the VeriFone Winnebago. And we'll ask questions. We'll ask questions in a very non-threatening manner." And he said, "We will come back to you with insights that nobody else can come back to you with." And I just resonated to that. I'm a little crazy myself. I resonated to that. And we got the Winnebago and we came up with a product that has in fact won a number of design contests. And the reason was: (a) we empowered the designers; (b) we listened to them; and (c) we trusted them. And they're

73

damned good. So in every way we won. And it was extraordinarily exciting.

John Kao has mentioned that our metaphor for our organization is a blueberry pancake. The reason I coined the metaphor is because we don't have any location that is the traditional headquarters. I was talking to somebody in New York and he said, "Where are your headquarters?" and I said, "I don't have them." He said, "Well, why not?" I said, "Let me ask you a question before I answer that question." He said, "Sure." I said, "Have you ever thought about what constitutes a headquarters?" He was really taken aback, because he didn't really expect that question. He just assumed that every company has a headquarters. And I said, "No, we don't assume anything. We challenge everything."

I said, "Let me tell you the classical definition of a headquarters in corporate America, or in corporate Germany, or in corporate Japan. That's where all the mucky-mucks are. There's the CEO in isolated splendor. You've got all the big vice-presidents in their offices, all next to each other. The wisdom is that because they're all next to each other, and their offices are next to each other, they talk to each other. And all these wonderful ideas flow. And Wall Street loves that garbage. And all the wisdom emanates from headquarters."

So it's kind of interesting that when I ran thirty commercial operations, I was the only corporate officer who was not based at headquarters. Headquarters was in Blue Bell, Pennsylvania. And I ran all operations out of Minneapolis. I had a lot of pressure on me to move, and I said, "Why do you want me to move?" I told the CEO,

> You've got all these guys sitting next to you, right? And they're supposed to talk to each other. Joe, wake up. The only time they talk to each other is when they send memos to each other. They don't talk to each other. I talk to you a lot more than they talk to you because I come to you. I'll come as often as you want, to pay homage to the wise men of the East, as I used to call them. I do what I have to do; you want me to pay homage, I'll pay homage. But leave me be.

While I'm running VeriFone there's no way I'm going to have all these trappings. So there is no headquarters because the CEO isn't based anywhere. I don't spend any more than 10 or 15 percent of my time physically in any one office. I'm not a complete gypsy, I do have a home. My home is in California, but my office is all over the place. You may say, "Yeah, well, you may be all over the place; but your staff is all in California." I say, "Hell, no." The staff is where they want to live.

For example, I've got my chief information officer, a fairly eclectic type of an individual, who chooses to live in Santa Fe, New Mexico. He's a very, very good person. I have nothing in Santa Fe but my chief information officer. I am very happy to have him in Santa Fe because he's extremely effective. I don't need to see him every day. I talk to him. I work with him. I e-mail with him. The vice president of human resources is in Dallas. The individual who runs development and manufacturing is in southern California.

We don't have a headquarters, and there's no reason to have a headquarters. My feeling is that if we in corporate America question these things a lot more, rather than just accepting things at face value, I think we'll all be a heck of a lot better off. So how do we run the place? We run the place as you've surmised by now as a virtual company. Everything is electronic.

Let me step back for a second. In this day and age a lot of people in large companies, small companies, and medium-sized companies, say, "Oh, yeah, we have e-mail." That's passé. E-mail, per se, is passé. One hundred percent utilization of e-mail is not passé. If you start to dig into the companies, you'll find out that it may be used in some ways or some groups may use it, but does everybody use it to the total exclusion of paper? I submit to you that you will not find another company that does it to the extent that we are doing it.

Now the fact of the matter is that there are tremendous advantages to being 100 percent electronic. There are some disadvantages too. Many people feel that imposes tremendous, almost self-induced, stress. Because when you are 100 percent on-line, you are expected to log on. There is no excuse. Either you log on or you are out. The fact of the matter is that you take the good with the bad. And so we have electronic mail and that's how we work. We recognize electronic signatures. All authorizations are done electronically; I sign nothing. I might get a capital authorization request for a $40 million plant in Shanghai, China, which we're in the process of constructing right now. There was not a single piece of paper generated. Usually there's a thick volume of paper and people signing and not signing as the case may be. We did it and we did it totally electronically.

These are some of the incredible internal benefits, but there are also some tremendous external benefits when you've got this kind of a mind-set. You notice I keep coming back to mind-set. Because I believe that, fundamentally, a corporate CEO has to really be 95 percent psychologist, 4 percent technologist, and 1 percent mildly crazy. Then you've got something going.

We had a situation where we got into a big dog fight in Greece. We were trying to close a major account, and the competition came in and said that we really didn't have the expertise that we were claiming we had. The sales representative didn't really know what to do. The prospect said, "If you really have that expertise, tell me where all these kinds of products are installed worldwide. If you can give me definitive proof, I'll go with you. Otherwise I'm told that you don't have this expertise."

Now you've got this situation. The salesperson is in Athens, Greece. It is Wednesday. The decision was going to be made in the next two weeks, but the tender was closing Friday night. The customer said, "Unless all this information is contained in the tender, I won't respond to the tender." The salesperson happened to be Dutch. He e-mailed to I_SALES, an e-mail alias that goes to everybody in the field organization worldwide. He said, "I've got this situation. Whoever has installations and can give me names of reference accounts and so forth, please respond. I need a response within twenty-four hours; otherwise I won't be able to respond to the tender."

He recounts that in the space of sixteen hours he had something like twenty-six responses. And what he did was a nice bit of marketing. Rather than transcribing the information, he just printed the screens and inserted them in the proposal. This is how we work. "Oh, by the way, Mr. Customer, here is your response. What is your next question?" I think it stands to reason that we won the account.

So we used the technology and used it to devastating effect. We have about one thousand of our customers on our e-mail system. You can imagine the competitive advantage when your customers know that any time they have an issue they can send you an e-mail, and they know they will get a response. What it does is tie your customers even closer to you. There are obviously many, many other things that we use e-mail for: small things, large things, ordering business cards, itineraries, anything and everything. The entire company runs that way. So that's who we are and that's how we do what we say we do.

I would like to end as I started, by talking about fostering an atmosphere of caring and creativity that drives people to really perform. We've launched a new initiative that we call VeriLife. Basically this program is a number of different initiatives focusing on the need for employees to balance work and family issues, while supporting VeriFone's goals to promulgate continuous change within the organization. In addition these programs try to preserve and expand on our corporate culture. I'll share two of them with you, but there are many others.

The first one we call VeriKid. The key element of VeriKid is the company serving as a catalyst to bring together the young children of VeriFone families around the world. What we are doing is making it easy for them to talk to each other via e-mail. We sponsor programs in the summertime enabling the young child of a VeriFoner in Germany to come to California. Our people do the matchmaking to see that the child can stay at the home of a VeriFoner in California, in the best type of situation. And we really try and help the children along those lines. We pay for the travel costs of the children. If that doesn't bind the people together and bring the VeriFone family even closer together, then I don't know what will.

Very often you might have vacation time and you might lose it. You're very busy and you can't take it, or whatever the situation may be. Or you just want to do someone a favor. This came up out of one of the focus groups that my staff and I run when we are at different locations, always talking to our people, always wanting to listen. One of the folks came to us and said,

> I've got a real problem. My spouse is very ill, and I have to take care of her. I can take so much vacation time and so much paid leave. And I know you will support me, but beyond a certain point in time, I cannot get paid. I understand that. I'm not complaining. However, out of that adversity I would like to see something in place so that when somebody else is faced with the same situation the company is there to help.

So we said, "OK, but what do you have in mind?" And he said, "What we have in mind is that an individual is free to voluntarily donate their vacation time to a bank, so to speak, of vacation hours."

And to take that particular case in point, heaven forbid, somebody's better half is not well, a child is not well, or some family emergency occurs, and they need to take an extended period of time off. Now they can dip into this vacation pool and take a much longer period of time off with pay than they might have been able to do otherwise. And you get into the situation where you say, "Well, how are you going to administer this?" And we don't claim to have all of the answers. We put a group of people together to try and work this out and put it in place. And we are hopeful, in fact, that this will become a reality.

The reaction from our VeriFone family when we announced these two programs was so overwhelming; I have to confess to you that it was an emotional experience for me in terms of the feedback I received via e-mail from people around the world. Because their fundamental comment was, "We say we should care about our people and I guess you're showing us that we do."

I would end, then, on a final note about human nature and the difficulty, the ongoing difficulty, of establishing an organizational design of the kind we have established, and keeping that design alive. To that end I am just going to show one slide that I brought with me. This is a slide I have of a poster in my office. I can't claim ownership because I didn't create it. It is an advertisement that some brilliant designer created, and that I picked up on. Let me try and indicate what the stripes say and then try to indicate the message behind the slide.

The slide has six blocks on it; the actual poster has twelve. So you've got twelve blocks, three across. In each of the first eleven blocks there is a golden retriever that is standing. On the twelfth block, on the extreme right-hand side, the golden retriever is sitting. In the first eleven blocks there is one word under the golden retriever and it says, "Sit." At the end of the twelfth block, when the doggone thing has finally sat down, it says, "Good dog." And the legend says, "Some messages have to be repeated a few times before they sink in." I would change that legend a bit and say, "All messages have to be repeated a few times before they sink in."

We are very mindful of that, and as we work to build our company, we will never forget what it takes to interact at a human level.

III. Design: New Institutional Agendas

Redesigning the National Endowment for the Arts
Jane Alexander

It is delightful to be here in Aspen in this grandest of natural designs, the Rocky Mountains.

Our subject today is new models of design in institutions, and I thought I might design my remarks around three institutions: the federal government, the National Endowment for the Arts (NEA), and its Design Program. As you know, all three of these institutions are in the process of reinvention and in the case of the latter two, very possibly elimination.

The National Endowment for the Arts has been a lightning rod for what ails many of our discontented citizens today and for the past five or six years: We're corrupting society with pornographic or homoerotic art. The arts are a frill or elitist. Since only the wealthy care about dance, theater, modern painting, and opera, let them pay for it all.

I try to counter these arguments: "The NEA has made 110,000 grants in its thirty years, and only forty have caused some people some problems," or "More people attend non-profit arts events in the U.S. than all professional sporting events combined—that's a pretty huge elite!" No matter what I say, the current antipathy toward public funding of the arts persists. It carries over to some of us in government, too. The other day I was in a store and the fellow behind the counter said to me, "Anyone ever tell you, you look like Jane Alexander?" I said, "Yep, they have." And he says, "Makes you mad, doesn't it?"

Well, what is going on? What is all this culture bashing and anti-visionary thinking? It comes with research and development in any field (the sciences, too). What is it all about? It didn't exist with our Founding Fathers:

> We, the People of the United States, in order to form a more perfect
> union, establish justice, insure domestic tranquility, provide for the
> common defence, promote the general welfare, and secure the blessings

of liberty to ourselves and our heirs, do ordain and establish this
Constitution . . .

That, of course, is the preamble to the Constitution, and it is the
blueprint for our representative and democratic government. The
framers of the Constitution created a flexible document that meets the
challenges of the centuries past and the centuries to come. By outlining
the characteristics of the role and responsibility of government, they
fashioned an institution that has grown in size and scope. Good
government has the common good in mind when it makes decisions
and creates and tests programs as solutions to social and economic
problems.

I believe that one of the essential roles of government resides in the
charge to "promote the general welfare of our fellow citizens" and
"secure the blessings of liberty." Design plays a key role in promoting
our welfare in several ways. First and foremost, design is about an
aesthetic. Tom Peters said last night that he may not be able to define it,
but I believe one can be educated to recognize excellence. And our
economic welfare benefits from excellence in design.

One of the fastest-growing occupations is design. In the two decades
between 1970 and 1990, the size of the labor force of designers grew by
156 percent. Architects have the highest median income among all
professional arts occupations, and designers are among the top five. It
might interest you to know that 1.3 million people make their living as
artists—more than agricultural workers, lawyers, or police. Commerce
is improved through better design of U.S. products, streamlining the
manufacturing process. And there is room for growth, particularly in
product design and all the new technologies. One of the mandates of the
Endowment is to help make our citizens, "masters of their technology,
and not its unthinking servants." This was written as part of our legislation thirty years ago, in 1965, when technology was the color TV, the
Mixmaster, and the room-sized computer.

Our environmental welfare depends upon design to provide a clean,
safe, and sustainable environment, making contributions in pollution-
control and recycling, as well as wise use of our natural resources, land,

and infrastructure. Yesterday I was giving out some Federal Design Achievement Awards in Golden, Colorado at the site of one of the winners, the Solar Energy Research Facility. It's alarming to realize that government funding for these kinds of projects may be pulled from the Department of Energy in the near future.

In education, design is a tool for identifying problems, analyzing information, developing critical thinking, envisioning options, and communicating solutions. When we teach our children design, we teach them the geometry of the world. We start babies off with blocks and Legos, and then in the first grade, we say, "OK, time to put away the toys now."

Every one of you here realizes just what it takes to be a good designer or architect—the comprehensive knowledge it takes to design a building, for example. An architect must have a sense of aesthetics, a command of physics, mathematics, geometry, and calculus. An architect knows about history, the sociology of the community, and the traditions of the built environment. An architect understands what dancers understand: how the body moves in space. An architect intuits what a composer knows: how harmony in composition works. Form and function marry in architecture as in all of the arts. A building is like a well-crafted drama, its component parts melding into a contiguous whole. A good building cuts across the disciplines and interacts with poetry and painting. It becomes a symbol as well as a useful thing. Think of how architecture inspired Georgia O'Keeffe to paint the New York nightscapes or one of America's great poems, "Brooklyn Bridge" by Hart Crane.

Students are starved today in much of their K–12 schooling. They need to see the relevance of school to their lives, where it fits in the overall blueprint of their lives. They need the knowledge and skills to make their way in the world, in their personal lives, and in their jobs. When they join the workforce in the new century, they will need to be self-directed learners. They will need to know how to identify problems and generate creative solutions and how to be flexible, independent, and collaborative. This is where design plays an important role. By virtue of its interdisciplinary nature, it provides contexts for teaching everything else, especially how to identify, redefine, and solve problems. It teaches how to analyze different types of information, understand the needs of others, the environment, the nature of materials, and so forth. The process moves from hypothetical situations to the real world.

Our children deserve the tools to meet the challenges of this new millennium. Design is elemental to that toolbox. We need to take advantage of our ordering nature as human beings, and the acuity of vision of our most imaginative students, to chart the planet's future. Commerce in the twenty-first century will depend upon industry doing a better job of design and manufacturing. Communities of the twenty-first century will rely upon planners doing a far better job of design. In fact, I cannot think of any aspect of modern life that would not benefit from the work you are doing. Just as design is fundamental to our children in the schools, it is basic to society at large. It can break down the physical and psychological barriers to full participation in society and open the way for a democratic and economic system that is truly inclusive.

Let me just mention one physical barrier that confronts me personally every day. The place where the National Endowment is headquartered, the Old Post Office, is old indeed. It was built over a century ago at the end of the Victorian era, and it inscribes that age on the modern world. The Old Post Office is a fort, a castle, a stately matron on Pennsylvania Avenue. There are a number of similar great grey boxes still standing around the country. From the outside it's impressive, but inside, in terms of today's needs for the NEA, it's a disaster.

A great central atrium, nine stories high, lets in natural light by which the postal workers of the nineteenth century sorted the mail. Offices are strung along corridors like beads on a chain. Communication and interaction among offices on different floors requires a real hike. You have to walk half a block to the nearest restroom. In that grand but anachronistic space, we attempt to accommodate new ways of thinking about business. Total quality management, team management, local networks, interactive systems—all of the tools of contemporary institutions—are made more difficult, simply by the building's design.

There's a little cast-iron spiral staircase outside my front door that leads directly to the floor below. Nobody seems to know its origins, and today it goes unused. Perhaps in the nineteenth century there was another office below my space that was important enough to make this shortcut necessary. In an imperfect building, it was the best design solution of the time. That spiral staircase is an apt metaphor for some of our most pressing problems. We cannot redesign the building, so we develop utilitarian solutions to meet our needs. Although as spiral staircases go it's aesthetically-pleasing enough, it is primarily functional, even though it no longer has a true function. It's a stairway to nowhere.

For too long, we in government and business have allowed design to be sort of like that spiral staircase. Design is useful when it solves a preexisting problem. The shortsightedness of such thinking is obvious. What we should really be doing is incorporating such principles at the outset, before we build our buildings, construct our programs, or establish our businesses. As I've traveled the length and breadth of America these past twenty months, I've often been appalled at the lousy architecture I see everywhere. People live as if they have no aesthetic sense at all, but I know that's not true, because you'll see a home with an awful, exposed-cinderblock foundation and two or three tulips planted in a weak-willed attempt at beautification.

As individuals, we make thousands of design decisions daily, from the selection of our clothes, to the path we choose to take to work, to the order we impose upon our schedule. The natural and human-made world around us reflects our drive to arrange and compose. Buildings and roads are systems, combinations of geometry and aesthetics. The revolution and rotation of the earth is measured and ordered into days and weeks and years. We design time down to the nanosecond. Our very speech is composed of patterns of phonemes and words. Every utterance involves a web of calculations that begins in the cradle, grows vastly more complex every day, yet works quickly and efficiently through that masterwork of all design: the human brain.

Because so much of our capacity for design is innate, we often take it for granted. And we have virtually ceased educating our children in aesthetics; it's rare to find any art-appreciation courses in our public schools today, much less painting instruction. We do not encourage enough those imaginative flights of our children nor help them with the discipline and skills necessary to turn that imagination into a creative result. What did Bucky Fuller say? "We are all born geniuses and get ungeniused with time."

Because we at the NEA believe that design is a strategic national resource, our Design Program, under the leadership of Samina Quraeshi, invests in projects which advance architecture, landscape architecture, urban design and planning, historic preservation, interior, industrial, product, graphic, costume, and fashion design. I love what Karl Lagerfeld said recently about Grunge: "They don't understand we're all part of the picture." The NEA also makes a strong commitment to design education, because while we all have an innate capacity, we need that structure of formal education programs to appreciate aesthetics and acquire the skills. It's the old form/function/vision paradigm.

Let me illustrate this with the story of Maya Lin, designer of the Vietnam Veterans Memorial in Washington, D.C. and many other public works. You may not know that the Arts Endowment held the design competition for the Vietnam Veterans Memorial, and twenty-year-old Maya Lin, a Yale undergraduate, submitted the chosen concept. When it was first submitted to an Arts Endowment panel, the arts experts who make our recommendations were wildly enthusiastic about her proposal for a simple black wall with a list of our fallen Americans.

Public reaction, however, was mixed. Many people felt it inappropriate or somehow disrespectful. Others suggested that something more traditional be accepted, and Maya Lin was challenged to incorporate realistic sculpture of three soldiers in her concept. She refused. This twenty-year-old girl refused to compromise her vision. And that vision is acute and overwhelming. If you walk west on the Mall in Washington, D.C. toward the Lincoln Memorial, you will come across a black gash in the earth. Two marble triangles meet in the middle, and on those long black slabs are the names of the dead—57,000 of them. Memorialized there are the men and women who served our country during the Vietnam War.

It is a quiet place, a sacred place, a place that spells out our sense of loss in each letter of every name on that wall. It has become a symbol of how we worked out the complex and controversial reactions we had to that war. The wall attracts millions of visitors each year. They touch the names. They make charcoal tracings of the names. They leave flowers and flags, remembrances. The Vietnam Veterans Memorial is a powerful demonstration of art as a healing force for a nation, for it speaks to our collective spirit. The wall is also a testament to the vision of one designer who understood that even a highly aesthetic object, a work of art, has a function.

The example of Maya Lin and other excellent artists we have aided in realizing their vision is only part of the story of the Design Program. In addition to awarding grants, we take a leadership role in identifying critical areas of need and, by initiating specific projects in response, seek to be a catalyst for change. For example, we fund the Mayors' Institute on City Design, which brings together mayors, designers, and urban planners to look at solutions for our cities. We fund the Federal Design Improvement Program, which solicits and rewards good plans, buildings, goods, and services throughout the federal government; Design for Housing, which emphasizes universal design; Your Town: Designing Its Future; and many other initiatives.

Our Design Program fulfills the Endowment's mission of fostering the excellence, diversity, and vitality of the arts in the United States and broadening public appreciation of the arts. And speaking of diversity, I loved what Tom Peters said last night about diversity and creativity in the business world. I've seen the same thing in the arts. Some of the most exciting work comes from the heritage of one culture or race bumping up against the contemporary environment it finds itself in here in the U.S.A.

For you here, the very nature of your design disciplines, with their dual facility for utility and vision, mirror directly the Endowment's dual mission. Design is first and foremost an art form, an expression providing vision, inspiration, and enlightenment to all Americans. We value the designer as artist at the National Endowment for the Arts.

As we near the millennium, rapid advances in technology are bringing about profound changes in the ways humans interact with one another and the environment. The resulting social, economic, and cultural changes, at both community and global levels, present challenges to designers to work across professional and geographic boundaries. Our Design Program is a unique national interdisciplinary resource for you all. Through grants, it encourages projects that foster effective communication and collaboration among the various design disciplines and allied fields. Through its leadership, it harnesses the power of your art now and in the future.

That is why it is essential that the federal government continue to maintain a role in support of design and all of the arts. The National Endowment for the Arts is here to serve the American people through the arts, to help create a better country, community by community, for the common good of all. But we are struggling to fulfill our mission. We have already lost 47 percent of our purchasing power since 1979. The budget is $167.4 million and falling, and yet the not-for-profit arts industry generates $3.4 billion back to the Federal Treasury in taxes! One NEA dollar leverages an average of $12 from other sources, an investment record that other agencies can't match. The Endowment is a remarkably good investment for the American taxpayer, who gives us 64¢ per year, the price of two postage stamps. And still there are forces in Congress that seek to eliminate the agency altogether. To balance the budget? Hardly. We're a drop in the multi-trillion-dollar budget, and a stimulant to the economy to boot. So, what's it about?

Picasso said: "Painting isn't an aesthetic operation; it's a form of magic designed as a mediator between this strange hostile world and us, a way of seizing power by giving form to our terrors as well as our desires."

That's scary—art as magic. But that's exactly what it is. It emanates from the deepest recesses of our brains and guts. We can't always tell how it got there or how it's going to manifest itself. The best art taps into our terrors and desires, not always pretty either. Think of *Guernica* or Goya or Gaudi. Art also seeks to define our relationships to others and space. At its most sublime, it speaks to a force far greater than ourselves. It is mystery and it is magic.

I think we have a lot of people in Congress today who fear the magic—who fear losing control. That's what art represents to them. It's part of my job to assure them that it's OK. We'll all be better off with more creativity and imagination in our lives, with more connection to community through art. We are, after all, spiritual beings, and what separates us from the animals is the ability to create symbols that have meaning for us—from the alphabet to the Bauhaus.

Symbol and metaphor are the artist's tools. They are the highest form of human creativity. I speak to Congressmen and Congresswomen daily about art, and my words often fall on deaf ears. I may not be able to succeed, but if we don't it will be government's loss—a profound loss, a dark ages for our U.S.A. All of you here will continue to persevere in your own grand designs, however, and always remember one true thing: throughout history, through time immemorial, artists will always have the last word.

Can We Count on Connectivity?
John Thackara

My talk is about the relationship between design and institutions. I want to make a couple of general points—mostly critical—about the agendas and policies of existing design institutions, and I'll take a quick kick at the idea of an American Design Center while I'm at it. And secondly, having criticised institutions in general and design institutions in particular, I'll try to justify the fact that I'm involved in setting up a new design institute myself. I'm sure there'll be time at the end for you to attack my double standards and inconsistencies!

If you add together design promotion bodies, professional and trade associations, and specialist colleges, there are the best part of one thousand organizations in the world that deal with design without actually designing anything. It's a moot point to me whether they add much of value to the design process. On the contrary, one could argue that by reinforcing design's "outsider status" they do more harm than good. Given the rate of change in the economy, technology, and in knowledge generally, it seems obvious to me that if you stand outside the processes of innovation and development—as most of these design institutions do—you're at best going to be left behind and at worst going to slow things down. In thermodynamics they call it *entropy* when a process becomes disengaged from its context; I think many design organizations are entropic.

There are one or two ethical and moral agendas to justify the existence of professional bodies, but I do think it is time to bury the whole idea of "design promotion" once and for all. Like muezzins calling to the faithful, a variety of design leaders have been chanting "good design means good business" for eons. Well-meaning Victorians in the U.K. started to worry about the design of British products in 1831.

In fact a cultural revolution in design promotion and its institutions is under way already. This became clear to me at a meeting I attended in Paris about eighteen months ago. The French design council, APCI, carried out a big survey among Europe's main design-promotion organizations, and the purpose of the Paris meeting was for them to meet and consider the results. I snuck in as a self-appointed expert on the subject. Normally, such affairs degenerate quickly into a wailing noise with designers and their "promoters" lamenting that horrid, narrow-minded industry still has not embraced design and that society as a whole is still failing to show "respect." Amazingly and wonderfully, this did not happen in Paris. One design promoter after another stood up to say (I characterize six hours of talk):

> We've been doing it wrong for decades. If industry has not embraced design, it is through our failure as communicators. Let us re-think our whole approach, and stop decrying the ignorance and moral turpitude of those we're supposed to help.

New strategies for design promotion around Europe were proposed in which the new role of a design "promoter" would be to connect different groups inside and outside the industry. The approach was called, without great originality, but with some accuracy, "Think globally, but act locally." Broad-band propaganda for design—magazines, design exhibitions, awards, schemes, and the like—was thought less useful than industry-specific research. The role of the design organization would be to bring industry and higher education together.

Advice projected at companies from outside is less effective than information delivered as part of a business relationship: managers rightly prefer to be treated as clients rather than as a child, a patient, or a sinner.

But these ideas still assume that you need design organizations at all— and it's arguable to me that the most successful design countries are the ones without design centers. Italy is one example; the U.S. is another. If you really needed a national design center, you would have one by now. In Europe we have dozens—hundreds—of professional and promotional organizations with the word *design* somewhere on their masthead, and, quite frankly, if they were all forcibly dissolved tomorrow, it's far from clear that anybody would notice, or care.

Look at Britain. The British government came within an ace of chopping the Design Council by 90 percent last year, and reaction to this brutal proposal ranged from disinterest to "Why keep it at all?" The only mystery to me is why they held their sword.

Be brutal about it: Where is the market for design promotion? Who needs it? I know designers think they need it, but where are the real customers? There are none. Mind you, this small detail has not prevented a variety of misguided people from proposing harebrained new organizations to ply just this trade. Japan, Taiwan, Korea, and Singapore between them have plans—some near realization—for about thirty new design centers. Several, bizzarely, are modeled on the British Design Council. Since 1990 two or three different groups have tabled plans to create a European Design Center, although the European Commission (an underrated organization) has been rightly dismissive of the idea.

I got into trouble earlier this year for attacking another proposal for a so-called World Design Council. I then discovered that the idea was actually an intiative of ICSID (International Council of Societies of Industrial Design) itself, and I can see many ICSID board members here in the tent today. But—and I do not mean to be personally impolite—I still think the idea is crazy. And the same goes for the concept of an American Design Center. If the words "design promotion" or "respect for the profession" or any combination thereof appear in its rationale, please, don't waste your money and your energy.

Design promotion is over. Delivering subsidized design consultancy to clients is over. Design exhibitions about design are over. Seminars with the title "Good Design Means Good Business" are over. "Good design" labels and accreditation schemes are over. Lecturing people about things they don't want or need to know, such as "the respect that is due to design," is over.

I mentioned that professional bodies—bearing in mind that a profession is an institution in itself—may have a role to play in today's messy, complicated world. Many professional organizations are indeed modernizing themselves in the realization that the old model of "professional practice" as the inward-looking protection of vested interest is inappro-

93

priate in today's new economy. Design's notorious inferiority complex—
"Nobody appreciates how wonderful we are"—is fading at last. Well, in
many places.

One reason is that quite large numbers of rather expert clients are
spending serious money on design in today's economy in Europe, in
America, and in Japan. We're no longer talking about, or as, a cottage
industry. We should realize that our tasks are to build on success, not to
spit in the wind.

Certainly, the design economy in Europe is in much better shape than
most observers expected after a truly dreadful four-coming-on-five-year
recession. A report we prepared discovered that the European design
industry turned over $9.5 billion (U.S.) in 1994, and is set to exceed
$14 billion annually by the year 2000. The size of the sector has been
obscured by its fragmentation (there are more than 8,000 design
consultancy firms in the E.U.) and by the fact that few client firms
account for design, whether purchased from consultants or carried out
in-house, as a discrete item.

But probably three hundred firms in Europe spend more than $1
million a year on design. Some, such as the bigger car companies, are
known to spend several times that amount. Design in big manufacturing
firms is becoming a capital-intensive activity, with very large sums
being spent on computer-aided design and model-making systems. As a
service industry, European design draws on more than 30,000 new
graduates each year. At any one time, more than 180,000 students are
enrolled in design courses in Europe.

And another counter-intuitive trend: Future growth in the design
consultancy market will come equally from the public and private
sectors—top five-hundred companies; an estimated 8 million small and
medium-sized firms; governments privatizing state-owned utilities; and
some three hundred different cities and regions competing for inward
investment. All of these will invest significant sums in design within
product development, branding, and communication programs.

As other speakers have already explained, computers and communication networks are a very large new market for design: Interactivity confronts the design industry with its biggest opportunity since industrialization. The Infobahnen (Internet)—computers and communication networks of almost limitless capacity—will draw on design to help develop products or "content" to match the stupendous scale of the investments being made in the infrastructure.

These different markets have a multiplicity of agendas, and within those the role and performance of design will vary almost infinitely. It follows that any concept of design promotion, research, or education which takes place outside those domains will be marginal at best. Only one model makes sense: a collaborative exploration of these new terrains by designers, clients, and users.

Let me say a quick word, before I come to the second part of my talk, about design and that dreaded word, "culture." Because if "the professions" are an institution, so too—with a vengeance—is "culture."

The 1980s saw an amazing coziness develop between design and all kinds of cultural institutions—books, exhibitions, biennials, television programs. But you have to ask: Was this a healthy rapprochement? Is design a proper subject for display and discussion in the context of art? Although everything from a spoon to a city is designed, it's a moot point to me—to put it mildly—whether design produces art or is an art itself. Is a Ford sedan art? A Pepsi Cola can? A Hoover vacuum cleaner? The "desktop" screen of an Apple Mac? What possible reason could there be for putting such objects in a cultural environment, still less for sponsoring the event?

Could money be the answer? Forty million French francs (about $6 million) was spent on a recent exhibition in Paris called *Design: Miroir du Siècle* (Design: Mirror of the Century). But the *Miroir* show was only the largest of a sequence of design exhibitions in Europe, Japan, and the U.S. which found their way into museums and galleries during the 1980s. In Japan alone, there were five hundred non-commercial design events in a single "design year" in 1989, the majority of which were sponsored. Indeed, a number of dedicated design museums were then built or are being planned.

Critics charge that design in museums is a pathological symptom of commercial interests polluting the cultural environment. And it is true that for many sponsors, the opportunity to engage in "cultural branding" of themselves as a design-led enterprise is attractive. This works well for firms wishing to present themselves to consumers as design-aware, such as car manufacturers or house builders; design as culture is also meat and drink for businesses selling directly to architects and designers, such as contract office furniture firms. Italian furniture companies in particular are masters at the art of displaying chairs and filing cabinets in museum contexts as *art*.

The thing is, I'm not actually worried about the effect of all this on art. I'm worried about the effect it has on design. I think Margaret Mead put it best when she said that "by making art a specially precious part of life, we have demoted it from being all of life." Until recently, design was "all of life," and despite the 1980s, it still lives down here in the real world. And I say, let's keep it that way.

This brings me to the third and final part of my talk today, or rather, to a question that may well have occurred to you by now: If design institutions—promotional ones, professional ones, or cultural ones—are such a bad thing, how come you're setting up yet another one in Amsterdam?

We call our project (its official name is the Netherlands Design Institute) a "think-and-do tank." This is because we are desperate to be part of the changes happening in design, not just commentators about them. Our job, if you will allow me to quote the mission statement, is "to identify new ways by which design may contribute to the economic and cultural vitality of the community." What we do is develop scenarios about the future; that's the thinking part. And then we undertake research projects to test the scenarios, the doing part.

The Institute's end-products are ideas, knowledge, and relationships. Now I know that sounds vague, but the point is that in helping designers prepare for the future, it would be absurd for us to pretend to know what will happen. Nobody knows that. But what we can do is figure out what the most important questions are going to be—from which mixtures of people and knowledge the answers might come—and then connect those questions, and those resources, to design. It's by making those connections that we try to help companies, designers, and researchers improve their capacity for innovation.

In practical terms, we have a full-time staff of eleven people who support the research, manage projects, and ensure that ideas and information are disseminated. On a day-to-day basis, our aim is simple; it is to make designers aware of, and ideally to get them involved in new issues or situations where innovation is possible.

We absolutely and resolutely do not base projects on traditional design "disciplines" such as graphic design, ceramics, illustration, or industrial design. All our work is organized around topics which allow designers to collaborate with others in the production of new knowledge. We organize ourselves around three programs. One is called "2D" or "Tomorrow's Literacies" and is all about the future of graphic communication. We try to immerse ourselves in the fast-changing communications environment and from there explore what new applications for the skills of graphic design, typography, and illustration are going to be needed.

We have a second group called "3D" or "Terra Nova" which is all about designing things in the new economy. The 3D team explores the increasingly blurred boundaries between pre-industrial craft production, small-scale manufacturing, and multinational corporations. The common denominator, so far, seems to be the consequences for design of the shrinking distance between the producers of objects and their consumers.

Our third and most advanced program is called "4D" and is mainly associated with Doors of Perception—the conference we started in 1993 on the design challenge of interactivity. It is our view that interactive multimedia and global networks confront design with its greatest challenge since industrialization. The opportunity is to develop products, or "content," to match the stupendous speed with which worldwide teledensity is increasing. The 4D program brings together social scientists, managers, designers, and others to consider the consequences of a network culture.

We're finding already, after two years on the go, that this vertical division between our three programmes—2D, 3D, and 4D—is being superceded horizontally by a number of social and cultural themes. We're finding that issues are much more powerful than organizational categories in connecting disciplines and interest groups together in our projects. I'll explain what I mean—and this is the end of my talk today—by talking about this year's Doors of Perception. We're focusing the whole project on the convergence of "info" and "eco" with the semi-unofficial byline that "The Killer-App Is Green."

You'll recall that we call the Institute a "think-and-do tank." Well it's easy to forget just how wide a gulf separates thinking and doing when it comes to environmental issues. For thirty years, scientists, think-tanks, governments, and global organizations have measured and analyzed the eco problem to death. They've produced a stream of ghastly projections. The result of all this is that "eco-gloom" is now a dark cloud in all our skies. But do we change our behavior? Do we? Heck!

In contrast to the gloom that pervades everything eco, life in the garden of information seems endlessly bright and sunny. You've all seen the Internet charts and connectivity projections that leap effortlessly upwards. Tens of millions will soon be connected, and we're told the information economy will grow exponentially forever.

There's just one small problem: Even the plugged-in among us will fry if the planet crashes! Global denial on this scale is a terrible frustration for environmental policy makers, but they lack the tools and under-standing to confront it. Their policies are effectively worthless until millions of people act—change the way they live in myriad small ways. But there's no sign of this happening fast enough or deep enough.

The experts talk about the "Factor 20 Scenario," a radical decrease in our absolute consumption of matter and energy within a generation, if we are to achieve a sustainable world. But it will only be achieved if a cultural shift of great magnitude takes place. No amount of legislation, and no technological fix, is going to save us. Factor 20 dematerializa-tion will only be achieved by intense, bottom-up economic and cultural creativity—stimulated by, but not relying on, new technologies, and involving a dynamic collaboration between businesses, experts, and the creative input of untold individuals.

So where on earth does anyone start? This is where our new project with Doors of Perception comes in. We're not proposing to redesign the whole planet and its metasystems in abstract. And we're certainly not getting into the business of global summits and grand plans to save the planet. But what we do think we can do is plant tiny seeds in the form of ideas and maybe also projects, that—who knows?—may produce one or two of the millions of answers we need to find.

Our theme, "Matter," reflects the shared interest of both communities in speeding up the flow of information in order to dematerialize activities that will otherwise devastate the planet's resources.

To identify the things that need fixing, we're learning about a process called *back-casting*, in which a scenario or "picture" is made of every-day life in a world which has achieved a Factor 20 balance. For example, "Ninety percent of food is now eaten within fifty kilometers of where it is produced." We then put these rather abstract scenarios into workshops—each one with fifteen mixed-discipline professionals—and ask them to work the consequences through. The workshops will take such ideas and turn them into "stories" that describe how life is organized in this new situation. Designers play an important role—using their planning skills to make the story coherent and their presentational skills to make it look persuasive.

The first group of workshops will focus on aspects of "Feedback." We tend to be confronted with the consequences of our actions for the health of the planet in a de-motivating combination of moral hectoring and abstract information. How might a combination of computer-graphic simulations and immersive media enhance our understanding of complex natural processes? How might information technologies refocus our attention on our bodies and on the earth? And in particular, how might scientists work with designers and communication experts to deliver this information in such a way that we relate to it personally?

Take the example of mapping systems such as GIS (Geographical

Information Systems) in relation to complex global processes. Many powerful simulations already run in laboratories; how might they be projected into society? The workshop will bring together social and scientific historians; psychologists expert on feedback and attitudinal change; designers and artists; and experts from the front-line of remote-sensing, GIS, and related computer simulations.

A second group of workshops is called "Caring for Matter." The ecological vision emphasizes the material presence of the planet itself; ubiquitous information creates a sense of immateriality and rootlessness. How might we use new information and communication tools to enhance our sense of, and responsibility for matter and place?

One workshop in this section will consider eco-tourism. The damaging impact of mass tourism is made worse by the tendency of modern travel to desensitize us to nature and culture: We move vacuously from duty-free to resort to beach—blind to the damage we may be causing. How might information technology enhance the concept of eco-tourism? This particular workshop actually starts at the the RMIT Winter School in Melbourne next month (July 1995).

Using telematics to replace environmentally damaging business travel and commuting is another topic. The idea sounds logical. But a much deeper understanding of the social and physical contexts of communication is needed if any impact on damaging mobility is to be made. Also in this section will be a workshop called "Electronic Songlines." In many cultures, shared values and laws on the environment are communicated through timeless stories, myths, and rituals. How might global information networks foster a better interaction between (highly misnamed) "developed" cultures and those wiser than our own? The workshop will develop scenarios for modern electronic storylines.

A group called "Eternally Yours" will run a workshop on the question of how industry might modify its reliance on the rapid innovation of short-life products. Should we design less desirability into hard products, or make hardware the "carrier" of infinitely mutable soft attributes? How might we communicate "time spent" as a value in products, not a cost?

Could we replace hard status symbols with soft ones, such as wisdom, friendship, care, entertainment, fantasy?

A third group of workshops looks at concepts of community. Take work and "telework" for example. Behind the rhetoric, the reality of much so-called teleworking is that it is unskilled and isolating. New telework concepts are needed that enhance social contact, which value both mental and physical skills, and which reevaluate the relationship between work and leisure. What are the main elements of this agenda? The workshop will focus on one or two live examples of new business concepts.

The concept of health is changing to encompass social and cultural factors as well as purely bodily ones. In a workshop on "tele-care," Francois Jegou and the Vormgevingsinstituut's Age Design team will consider the consequences of a virtualization of social relationships. Positive connotations—such as new social connections—may easily be cancelled out by negative ones, such as increased social isolation. How might *telematics* improve the social connectedness of those, such as old people, whom industrial society—let alone *informatic* society—has isolated? How may informatics alter current models of "social service"? The workshop will focus on a specific telematic application for old people.

In addition to these workshops, we will have talks and a conference on key themes. One of these is the concept of "collective intelligence." Information technology is delivering powerful new tools which make it possible for people to communicate with each other via machines, not just one-to-one, but also many-to-many. How real is the prospect that these tools will foster the collective intelligence—not just the exchange of data—that we will need to achieve a sustainable future?

Another topic is about "the mental and the material." In designing information networks bigger than ourselves, have we made ourselves blind to the vital signs that tell us about the health of the planet? Have we forgotten the fact that human intelligence is bound up with having a body and that our bodies can only exist as part of a planetary eco-system?

I've tried to emphasize that by *connectivity* I mean connectivity between people, between companies, and between ideas —not just the use of the Internet in a technical sense. But we are also using the new communication networks themselves to make this happen. Our Doors team includes some pretty experienced pioneers in the use of the Internet, the World Wide Web, videoconferencing, MOOs, and other media environments, which have the capacity to engage individuals around the world.

We already used multimedia to disseminate the results of our first Doors of Perception conference with a CD-ROM, and the results of the second conference were put onto the World Wide Web. But this time we're trying to use the Internet not just as an output device, but as a bridge between, and among, the participants. So we're building something called DOME , which stands for "Doors on Matter, Knowledge, Environment." The idea is to to give each workshop group its own space on our Web site and put in there questions, texts, links to other sites, booklists and so on—all the input you need to get a workshop started. Then attached to that static stuff we're building a discussion space where the participants can talk to each other about this material—basically by sending in e-mail messages that the system lodges automatically near the document in question.

Quite frankly I have no idea if this technical side of the process will add a little or a lot to the overall Doors program. But the point is that we're getting our fingers "virtually" dirty and are not just talking about the technology in the abstract. So even if this DOME experiment is a total disaster—and I don't think it will be, because we have some brilliant people from Amsterdam's hacker fraternity in our group—but even if it fails, we'll learn a lot about the difference between rhetoric and practice when it comes to the new technology.

So, to conclude: Can we count on connectivity? I said some rather harsh things at the beginning about the limited value of design organizations and then went on to tell you about our own organization in Holland. I should say that many intelligent and skilled people work in these design organizations; so if we can indeed steer their agendas away from preaching and towards connecting, the human infrastructure exists to do a lot of good.

I do know that in our experiments with Doors of Perception, the combination of thinking about a subject critically and showing the best examples does catch the imagination and enthusiasm of many people outside the design world. The lesson we've learned is that if you confront people with well-thought-out questions; if you bring different disciplines together to deal with them; if you create a good context in which to address them; if you are systematic and creative in disseminating the results; and if, at all costs, you avoid telling people what to think about design; well, if you do all that, people will respond very warmly.

IV. Design and Business: New Connections

Transform: Reinventing Industries Through Strategic Design Planning
Larry Keeley

I want to talk to you about three basic things. I'm calling the overall focus of the show "transform," the idea of changing things substantively from one thing to another like a caterpillar, which is a basically ugly, wormlike thing changing into a butterfly, which is a lovely thing that has a great deal more mobility and a great deal more range of movement and freedom in the world.

That's a leitmotif for the kinds of things that I think are important in design, important in business, and important in the special nexus between design and business that I thought I'd talk about today. The other thing that I want to do is to talk about how the design field itself is going through a massive, permanent transformation. I'm going to talk about how design, unbeknownst to most of us, has been permanently transformed. I'm going to spend some time talking about the design business as a business. Throughout this conference we've heard mostly from nondesigners talking about the wonderful relationship that should exist between design and business. Few people have talked to us about the design business as a business. Warning: This portion of my presentation will be depressing. OK?

The second thing I'm going to talk about is what big businesses really need from design beyond the evangelical sort of characterization of those ideal relationships that should ultimately evolve between design and big business. What I'm going to talk to you about is what they are asking for right now. Not necessarily from designers but from anybody. And that is really interesting and exciting.

And then the third thing I'm going to try to do, and this will be fairly perilous in the brief amount of time that I have to spend with you, is to try to talk about the most exciting relationship between the sort of switched-on designers and those emerging business needs that big business has. So that's my game plan and we'll see how it goes.

I'm unveiling a bunch of stuff here that we've never unveiled before in the hopes that it will be helpful to you. We do call this, by the way, *reinventing industry*. Reinvention is a big word, for those of you that aren't connected into the emerging business theory of the business schools. Reinvention is a bigger deal that re-engineering. It's what C. K. Prahalad and Gary Hamel talk about these days, and it's something that you'll hear more about as we go on. What I particularly like about the word *transform* instead of *reinvention* is that it has *form* in the word. And so it begins to immediately, right there, bespeak the role for designers.

Let me begin with a metaphor that connects us back to an older time. At the close of the eighteenth century, as some of you know, there was an emergence of automated looms. It was the first time ever in history that people could produce large amounts of fabric at low cost. The automated loom was a huge transformation in the history of the production of garments. What did it do? It made fabric available to a great many more people at much lower prevailing prices. But it did come at a cost.

The cost was in two forms. One is that the average quality of the fabric was much lower than that of fabric made on hand-looms by master craftsmen. The other was that all the people who were hand-loomers, the weavers of the day, were, of course, mortified by the advent of a new technology which was essentially throwing them all out of work. They became the famous Luddites that rioted and revolted against technology in general, trying to burn down those factories and destroy those looms and protest against this permanent transformation of their industry and their profession.

The Luddites did something that is not well understood. They objected to technology, but what they were also objecting to, and this is very important for designers, is the lower prevailing level of quality of the fabric that was produced in that time. They said, "It's not good enough. It's terrible stuff." Now here's the thing, and there's really no way to escape this, the exact same phenomenon is happening in the design field, right now, globally, everywhere. And we are not able to understand in the abstract or haven't spent the time to understand, the implications of this massive and permanent transformation.

Here at the Aspen design conference, I've been very fascinated by the undertow I'm picking up about technology. More than at most design conferences there are people here who are quite angry about, and resistant to technology. Harking back to the Luddites, right? But ladies and gentlemen, the Luddites lost, the Luddites lost permanently. Why? Because basically people wanted to have fabric in abundance at low cost, and because the technologists worked relentlessly to raise the quality to the accepted level that people would really like to have. The Luddites, in the main, did not accept a change, learn to embrace it, and learn to adapt. This is a message that will be heard again, I think, as events wear on this morning.

In keeping with the cinematic leitmotif that runs through the conference, I've decided to use some slides to give you a sense of the major themes I'm trying to introduce. This first major theme is the notion of change; change to the design professions and change that affects designers. So here I've borrowed from *Modern Times*, and the indignities that Charlie Chaplin suffered, in order to suggest the beginnings of this transformation that we as designers are going to experience firsthand.

I'm going to bore you with a few statistics and analytics. Most people haven't bothered to look at the pattern of the analytics. So let me just explain it to you. The top slope, which is rising at a fairly rapid clip, is the cost of entry for a professional designer in a four-year education. It continues to rise at the alarming rate that all sorts of higher education is rising at. The bottom curve, somewhat flatter, is the price of vocational-technical training. It's the price of learning how to use Quark and a Macintosh and a few things like that. Notice that slope. Quite different. There's a gap. An analyst would tell you this is a problem. They would tell you that that's going to have an impact over time. But wait, there's more.

In the United States, according to the best available estimates—and it's very hard to get good data—there are between 400 and 600 four-year graphic design schools. There are certainly over 600 schools that issue design degrees of one type or another. These folks, in the U.S. alone, produce an estimated 50,000 graduates annually that have some sort of certificate suitable for framing that says they can do design. OK?

The same phenomenon going on in Europe. In the European community, 180,000 students are studying design full time. Whoa. Each year, 30,000 graduate as designers. There are 9,000 design graduates annually in Britain alone, going into a country that basically, as best we can tell, only employs 10,000 to 11,000 designers.

To our friends who are listening in Japan, I can say that the same pattern exists there as well. The Japanese have committed to having a design school in each and every prefecture of Japan, because they know that it's a strategically important capability for the country. And that is producing the same abundance of supply of designers that we're seeing elsewhere in the world.

So the critical question is, can the industry begin to absorb all this incoming talent? Well, let's look at the best available projections that we're able to glean. What you find out is that there is a draw-up in the demand for designers, but it happens to be at a rate of about 1.39 percent a year. Not a very good rate of growth. And why is that? Because design, like every other part of people's work these days, is increasingly becoming automated to the degree that we're able to produce more stuff with fewer people at lower cost at higher speed all the time because that's the norm. That's what's going on. You really don't need to hire as many new warm bodies as you used to, because you can do more with the warm bodies you already got.

Now let's overlay the demand curve and the supply curve. Yikes! What you find out is that there's a large gap and it's growing and it's scary. Look, ladies and gentlemen, what this inextricably says—there's no way around this—is that the economics of the field are going to change. I told you this was going to be the depressing part. Now it's going to get worse because it's not just the warm bodies, it's also the things that we buy and the things that we do.

We have all of us loved to buy Macintoshes, laser printers, and the software systems that we need. If you're a designer, or even if you're just a regular person with no certificate suitable for framing, you buy yourself a Macintosh, a CD-ROM, a modem, and a decent printer plus Quark and a few other things; and you can reliably do many of the things that people are trying to do in the design profession. The software folks have, of course, loved catching up with that as well. So now, ignoring the huge number of people that are trying to enter into the field, and looking at the underlying costs of what is we produce and the cost of entry—now I'm looking at capital equipment costs—what does is cost to start up and be in business?

And look at that curve over just a very few number of years. The first year there that you can't see is 1984 and the dotted line is now 1995. What we're seeing is an enormous drop in the cost, the upper curve, of the equipment we need to be in the design field. But the bottom curve is even weirder and wickeder. The bottom curve is the variable cost for a job. That's what it costs you to do the typesetting.

Remember in the old days we all used to have to do thousands of dollars of typesetting to get comps to look halfway decent? Not any more, baby. With labor controls, models, factories, automated accounting techniques, and the ability to produce reliable and reasonably effective comps at the desktop, the variable cost for being in a design profession for a project is close to zero.

Now what does this mean in practical terms? It means many things, like this. Here's a precursor. In San Francisco they say that the going rate to produce the entire graphic package for an audio CD-ROM is seventy-five bucks. Now, why is that? Well it's because the designers are already there. They're living there, nice place to live. They've already got the Macintosh; they can do the stuff; and maybe they're going to get to meet Sting at the wrap party. Right? So, "Seventy-five bucks? Good, fine, no problem. It's just kind of a lark. I'll do it in my spare time." That's the way people look at it. And those are the kinds of things that undermine the economics of the design profession.

And wait, it still gets worse. How many of us have computers now? Well, the latest statistics say that computers have penetrated the average graphic design office in 96 percent of the cases. That's up from 94 percent only two years ago. Now some of the offices, of course, are just using it for accounting, but increasingly people are using it to aid in the design production.

Now let's look beyond the warm bodies and the technology at the software that comes along. And this is where it really starts to get frightening. You go to any Egghead, or even a discount catalogue for software, and you can find all these kinds of software. The one on the left offers up everything that you need to publish newsletters. The one in the middle puts you in the advertising business. The one on the right makes you a graphic designer—posters, goofus cards, whatever you want. And if you read the details, the small print on the packages, it offers up, "You can be a designer, copywriter, layout artist all rolled into one." And that's available for eighty-nine dollars. OK? There you go.

Now, of course, these are trivially stupid programs, and they do not an Ivan Chermayeff make. You all know that, and I know that as well, but again they change the underlying economics of the field. And it's going to get to be a very pervasive phenomenon. For those of you who are architects, it is very possible to go out and get 3-D Home Architect, which will allow you to design your own home in the privacy of your own home with your Macintosh. You can do the same thing with the landscape architecture tool that they offer as well. So these are the kinds of things that ultimately start to change the nature of the business.

So that was last decade. Let's look forward to next decade, because that's when it's going to get deeply grim. And to do that I'm going to give you the example of Seaside, Florida. Seaside, Florida, is this excruciatingly charming community that's been created out of a series of rules by architects Elizabeth Plater-Zyberk and Andres Duany. A very important case. It's the way to create rules that allow people to make an interesting sense of place. And, of course, the idea of Seaside, Florida, is to create themes and variations. And many of the best architects in the world have come and tried to live within the rules of Seaside, Florida, making it indeed seductive. So now it's this hugely successful financial result and a very popular honeymoon community.

It has things like this: By law, everybody in Seaside has to have a front porch. And everybody in Seaside has to have a picket fence. But your picket fence has to be different from everybody else's picket fence. And so the idea is that if you have a front porch you might sit on it and be there while people are wandering by admiring your picket fence, to say, "Very nice picket fence." You're on the front porch. You're on the sidewalk. You talk to each other. Hey, you've got a community. That's the idea. This is what we call a rule-based system. And rule-based systems are going to be huge, ladies and gentlemen, not just big, huge.

Rule-based systems are starting to come in a huge way. And this is the final way that technology and software systems will permanently transform the design field. And it's already happening.

Some of you might be familiar with the extraordinary work that Bill Mitchell has done in his role as the chief of the architecture department at Harvard. He's now at MIT, and he's running an architectural program there; and most of you probably aren't aware of this but the Media Lab technically reports in through Bill Mitchell. That's how it's structured at MIT. This is the work that Bill Mitchell did while he was at Harvard. He created *the logic of architecture*, software systems that allow you to create designs from Palladio, from Frank Lloyd Wright, and from Corbu, which is a neat trick because all three of them are dead.

And that is an interesting phenomenon. You can have Corbu design your home or your office as you see fit with the rule-based systems that have been created there at Harvard. That was done on mainframes with really exotic technology, but Billy Buck in Seattle is getting in on the game, too. At any nearby software store you can go, right now, and buy Microsoft home software that gives you the Frank Lloyd Wright design style. You can learn about Frank Lloyd Wright, and you can see many of the pieces; but, ladies and gentlemen, there's a part of the software program that allows you to design using Frank Lloyd Wright building blocks, sort of like a Lego set with style.

Now let's turn the time machine forward a little bit and ask ourselves what we're likely to get. And this is what Doblin Group projects we will see before too long. This is the Michael Graves lawnmower construction kit. It doesn't matter, you can make anything with it. You can make a tea kettle or you can make a building. It wouldn't matter. You probably can't tell, but in the small print of the pull-down menu you can change it, folks. Woody Pirtle, Philippe Starck, Michael Vanderbyle, or April

Greiman. And, ladies and gentlemen, for those of you who have made it already in the design field—and there are a few of you in the room—this is news you can use. Where's Saul? Where's Ivan? You guys have really got to get to know some good software programmers, because this is the way you guys can personally get, in the words of our esteemed Tom Peters, "filthy stinking rich."

There are ways we can make the software systems automatically control the royalty system. So that when anybody buys this lawn mower the Net automatically hits your bank account in the right way.

I'm making light of it, but, ladies and gentlemen, this is going to happen. There's absolutely no technological resistance. There are no technological breakthroughs that are needed to make this possible. It is entirely possible right now and what does it do? It allows the superstars of the design field to continue to be superstars. It has enormous implications for those of you who are young and just entering a profession.

Let's now have just a moment of the audience-participation portion of the show. How many of you as either professional designers or educators are, in fact, producers of design? Raise your hands. That's what I surmised, a lot.

How many of you are sort of like clients, consumers of design? A very small number, just what I expected. So those of you that are designers, I hope you were paying attention. If you're trolling for business, you've got a small target.

Last audience-participation question: How many of you are under thirty? My message for those of you who are is particularly pertinent. It's going to have a big impact on the design field during your lifetime.

Now let's step back from this a bit. Let's try and figure out, beyond the design profession, what does it mean? Where does it come from? Why is this different now than it's ever been in the past? The reason it's different is because of the rate of change in information-processing in the world. Companies like Texas Instruments, Motorola, and Intel are producing enormous numbers of semiconductors that are constantly increasing the MIPS power of the world, the Means of Instruction Per Second that are able to be processed in some powerful way.

And, of course, lots of clones are being produced at ever-faster rates and at ever-lower prices. What that is doing is making for an incredible rate of change. That's what this graph is. This graph has a very interesting time line. Way back here on the left side of the graph is like the beginning of life on the planet earth. And the vertical line is arbitrarily defined as now. And what we've done is we've gone through a relatively stable rate of change and suddenly it's gone up and up a higher slope.

There's a great man named Danny Hillis, a sort of pioneer in computing architecture, who talked about this. He said, "You really can't really make sense of this unless you put it on a logarithmic scale, like engineers like to do." What you do when you put it on a log scale is you find out that the rate of change was largely flat for the first several billion years of the planet's existence and then very recently it has gone through a big increase. And Danny Hillis asks an important question, "When in the course of human events have we had a similar rate of change?"

Anybody got any ideas when we've had a similar rate of change in information-processing? Beginning of the universe? The Renaissance? Yes, I often get those kinds of responses. The printing press. The Renaissance. People always think that it had something to do with human beings. Wrong. False. No. What it had to do with was the beginning of DNA itself. That's the first time in the history of the world that there was an increase in processing speed and information power similar to what exists right now. And this, ladies and gentleman, is not just a change in the speed of the world, not just a change in the computing power of the world. This is a change in the nature of the world. This is a permanent transformation.

So with that I'll move into the second portion of my gig with you today wherein I talk about where breakthroughs come from because, ladies and gentlemen, that's what businesses want. They want really cool breakthroughs. They are sick to death of having ordinary products like the type that Tom Peters was making fun of. They really do want breakthroughs; they just don't know where they're going to come from.

So here with a floppy-like disk of back-to-the-future is the second portion of the remarks I wanted to share with you, which focuses on where breakthroughs come from. What we see increasingly are surface-level manifestations of the people who are increasingly using this information power. Here in this really bad slide, you see a guy trying to do molecular engineering of new compounds, being aided both by a physical model and some virtual-reality headphones that are visual aids that allow him to process the information as he's manipulating it in three dimensions.

That is a typical example of the new kinds of ways that allow all of us to invent things at a faster rate than ever before. Invention. More break-throughs coming faster than ever before. Really interesting and excit-ing. And really exciting and important for designers. What that adds up to is a whole bunch of things that change in a nonlinear fashion. Things that came from no place and permanently changed the world. Let's look back ten years and ask about those things that came from no place and permanently changed the world.

The Apple Macintosh is only eleven years old. It's the computer for the rest of us, remember. It's the thing that allowed an enormous amount of computing power to be in the hands of the average person with no special training.

Battletech is a simulation gaming system that allowed people to have fun blowing things up, I guess. But it went from zero to $300 million in revenue in three years. Every time I talk about this in Japan it's really amusing because people immediately pull out their Battletech member-ship cards. It's really hugely popular over there.

Canon laser engines, another breakthrough product. Totally destroyed the world of xerography as Xerox knew it for many, many decades, and, in fact, has managed to achieve 97.5 percent global share. All those little replaceable, recyclable engines that you put in your laser printer. They're all made by Canon even if they're branded Apple or Hewlett-Packard. An extraordinary achievement. CNN, Motorola, Microtech—these are all things that have emerged in recent times.

Let's take a look at a couple of them. CNN is almost sixteen years old. CNN probably couldn't have existed without things like the Macintosh, the really small portable satellite uplinks that you can get from Sony at low cost, or really reasonably good low-cost color video cameras like people are using in the audience. And this changed the economics of the field. Let me put it into stark perspective for you. CNN was able to produce twenty-four hours of global news every day for less than the cost of what CBS was spending to put on one hour of news a day in North America alone.

Now if you're CBS and you're sitting there and you see this entirely new thing come along at a cost structure that you simply cannot touch, it's like the game is over. You've got to figure out what the heck to do. You've got Dan Rather. You've got Connie Chung. They've got these exorbitant salaries. And CNN's got these no-name people out there reading the news. It's not that hard; anybody can read.

The laser printer, it's a remarkable science. We worked years ago with Xerox. At Xerox they thought the copier engine was a holy artifact, a religious thing. They could never imagine that regular, mere mortals would be mucking with it. And the folks at Canon said, "Well, we just cannot begin to compete against Xerox's service network." So what did they do? They rethought it entirely. They said, "Let's make the engine replaceable." Absolutely miraculous achievement, conceptually, in terms of engineering, and in terms of execution. And it enabled the entire design field to have really low-cost, very reliable ways to produce comps. Now if you go into the world of strategy and you ask yourself, "OK, what do people have to say about this?" You get the kinds of ordinary answers that Tom Peters did a fine job of describing. Igor Antrop wrote the world's best-selling book on strategy. It's like twenty-two years old. It's ancient history. There are Michael Porter's books on strategy—the bible of strategy according to the Harvard Business School.

And, of course, in the course of human events the consultants have figured out how to give us a whole bouillabaisse of buzzwords which corporations are rushing to adopt at a serious pace. But they're all being TQMed and QFDed and they've got their strategic intent, they got their core competencies all noodled. They're doing everything they can to be multidisciplinary and to have the kinds of team tools and cross-functional teams that Michael Schrage will help them to facilitate. And they just think they're supposed to be just humming on down the road, but the results are, in fact, very deeply depressing.

There is this focus on speed which everybody else has addressed and about which John Thackara did a wonderful job of saying, "Oh, isn't that dreary?" But basically if you don't have a big idea, you better execute the idea you've got with some startling speed.

I want to take a moment to talk about how the conventional wisdom looks at things like the computing world. And I'm just going to take a moment to show you where weird and stupid products come from. All around the world there are lots of analysts trying to figure out, "OK, what's the future of computing?" They all know from Nicky Negroponte and his colleagues at the Media Lab that, basically, the name of the game is that they're going to get computing and content and communications to converge—the CCC model everybody's been talking about. But if you go to the best analysts in the field—and I stole this really bad graphic from them so don't blame me—they say the first phase of the computing revolution would be to get the infrastructure in place. So everyone buys their computers, and they buy their network, and so forth.

And the second phase would be enabling technology. That's when we're going to get the Internet really humming and we'll all have a modem for the highspeed T1 link from the communications companies where they have wireless connections and all that good stuff.

And phase three is going be where content comes to the fore. We start really doing these neat things. And they talk about the convergence and they produce models like this one. There's content and there's computing and there's distribution and there are companies that have core competencies and all those things. And boy, baby, look at that big black hole in the middle. That's where the mother lode is. That's where the gold is buried. That's where we've really got to hit it with a big hammer. And Apple, thinking about this many, many years ago, said, "Yeah baby, we

117

can drop something right on that; and that's an Apple Newton." Right? Right. Apple Newton is out there, you can get one. Hasn't been really much of a stocking-stuffer item, but they are out there in abundance. And there were many people out there who were early adopters at $700–800 and are still trying to figure out what to get it to do. When you think about these sort of technological imperatives, how things are going to come from someplace and be terrific, you tend to produce things like this that do not generally live up to their expectations.

This is the one I love. This is a huge billboard currently at O'Hare Airport. It's advertising OFC warp, or if you're like those nuns in Europe, "OFC *varp*." And if I were standing in the airport I would only come up to about here. This is a big woman, right? And she's got a big idea, and I know you can't read it so I'll read it to you. It says, "I can't believe you're running a video; you've got the Internet going; and you're faxing."

I can't believe anybody would think that's a really good idea. It's just another example of technology facing uncertain aims and purposes. Believe me, if you've peeked under the covers you understand that's a really nifty piece of technological knitting. And you can understand why some of those people are confused enough to want to advertise about it.

Other products that come from this same sort of thing are things like Sharp Wizard. This is a product that is truly ahead of its time. Only in about thirty- or maybe fifty-million years will human beings evolve fingers small enough to actually use that keyboard.

And this is one that we've created to sort of highlight a merging alliance between Hewlett-Packard, the manufacturer of printers, and Hoover, the manufacturer of vacuum cleaners. We call this the Print Vac. It's a terrific device. It's got all kinds of really important features. For instance, it will give you a continuous printout of everything you're sucking up around the living room. It's got the directional-turn indicators on the front so that if somebody's there reading the *New York Times* they can tell if you're coming into the neighborhood and they can lift their feet up at the right time. It's got the odometer so it can tell you just how many square feet of carpet you've sucked up, and it's got the second power indicator so that you can know whether or not your device is performing within the rated norms. This is the kind of thing that we think is really sort of unfortunate. But it appears all the time as this world converges.

And, of course, for designers and for the businesspeople who employ them, here are three enduring questions. Easy to ask, hard to answer. Three important questions that shape the way design should work. What matters to people? What's next? And how do you know? Those questions really make the difference and lead to breakthroughs. Those are the questions I want to talk to you about. When you do them right, you're able to connect things and people in richer, better, and more resonant ways.

Now things, artifacts, are like products or messages or even information systems. They come equipped with features and capabilities. And human beings, my favorite species, come equipped with their own set of capabilities. They've got values and beliefs and assumptions. They all have language and they all have skills. And the important thing is to find a resonant connection between them. And when you do it right, you do it in ways that change the relationship. And I'll take you through an escalating series of examples from simple to more complex.

Tucker Viemeister and his colleagues at Smart Design revolutionized the world of lower-cost kitchen utensils recently in a case that they did for Oxo, called Goodgrips. Most of you are probably already owners of Goodgrips, the terrific little peelers that were designed to be optimal for people who had arthritis. So they've got big fat rubber handles with little flanges where your thumb and forefinger will touch so that they're easy to grip and hold even if you're peeling carrots underwater. And a marvelous thing happened by designing a line of kitchen utensils for handicapped people. The Oxo folks managed to make a breakthrough, runaway success that allowed them to create better kitchen utensils for everybody.

That was a typical example of how, if you find the right connection between artifacts and human beings, it works out rather nicely. That whole field, by the way, of kitchen utensils has been very permanently changed with twenty or thirty other guys trying to pile on lately with the big fat handles, the nice rubbery grips, and so forth. One of the questions one would have is, "What could or should Oxo have done to protect its intellectual property along the way?"

Another case that shows this general pattern is the rather horribly brand-named phenomenon of the ASU Roll Aboard Travel Pro Luggage System—that's what it's technically called. It was originally designed almost fifteen years ago for people who travel a lot, namely pilots and stewardesses. And it was designed, not to be piece of luggage, but to be a travel system. It tried to understand the full sequence of indignities that people suffer as they go from one place to another and to optimize for all of those indignities.

So among the things that it allowed you to do was to pack your clothes conveniently, to allow them to be folded up compactly without wrinkling them beyond remembrance so that when you took them out there was some chance that they would look vaguely like clothes. It was designed to fit into a system that was quite aware of what the FAA regulations were for under-seat storage or overhead bins, and it was designed to have very robust handles and really good bearings and really good wheels, so that if you were not a very muscular person you could carry it along routinely as you went through an airport or into a taxi or bus or whatever. Finally, it had a system of hanging straps so that you could cascade additional bags over the top.

It was a simple idea. It's quite old now, twelve to fifteen years old. And the folks at AFU just did a fine job for the first ten years selling it to stewardesses and pilots who were fond of the system and bought it in large numbers. Finally somebody got there and said, "Whoa baby, there's a lot of people besides pilots and stewardesses that travel a lot." And so they finally learned to market it to the rest of us who go through all these same indignities. So now all of a sudden Boeing knows that it needs to be designing aircraft, more specifically, to anticipate and accommodate this kind of luggage system. That's the kind of ramification that these things have, the beginnings of something I'm going to talk to you about more deeply in a few moments.

120

Now I'm not one, personally, to ever give anybody in the automotive industry credit for doing anything right, because basically I think its populated by people who never grew up much past the age of twelve and really like things to go "Vroom, vroom," a lot. But the Dodge Ram truck is one of the rare examples of a project where somebody did anthropological research and tried to figure out what it was they ought to do to make a truck really accommodate its patterns of use. What did they do? They simply learned that an enormous number of people run small contracting businesses out of their trucks. So it's the global headquarters of the contracting business. And they said to themselves, "What could we do to accommodate that?" And they did a couple of fairly trivial simple things.

One of the things that they did, for instance, is include two, not one, but two, cigarette lighters in the ashtray. Not because they think contractors smoke twice as much as anybody else, but because they knew the contractors would want one power take-off for the cellular phone and another power take-off for the laptop computer. And similarly they created a really fat pull down thing in the middle that is designed to accommodate the cellular phone, has a place for the notebook computer, and it locks when it's put up on top. Behind the cab-seat they put these changeable kinds of shelving systems that can be adapted in one way if you happen to be a plumbing contractor and in a different way if you happen to be a woodworker or a metalworker, or whatever. And, of course, if you happen to live in Los Angeles or in Dallas you can put in the gun rack factor, which is very handy.

And so now I want to get to a fairly sophisticated example of how breakthroughs in design are beginning to happen. And for this I'm going to elaborate a little bit on the co-construction of innovations or the ways in which new breakthroughs tend to be coming from multiple dimensions all at once.

Most people in the United States used to believe, not that long ago— sort of the 1950s residue that lasted until the middle 1980s—that they would be employed by largely one or two employers all of their lives who would take care of their financial needs, and would vest them in some sort of a retirement program so that they would have the right kind of pension at the time that they needed to retire. And whatever their pension didn't cover, of course, the federal government would cover in Social Security and Medicare payments.

So that was the basic American idea. The employer takes care of you while you're able to work. The government takes care of you thereafter. It was a noble idea. It was a sweet idea, but, of course, it's been completely blown to bits in the last few years. We no longer believe this; we no longer believe the government will be able to take care of us. We certainly don't believe that the employer is any more loyal to us than he or she was historically, and probably less so. So we are all seeking new kinds of resilience. We are all seeking new ways to buffer ourselves against the vagaries of modern times and the uncertainty of our economic existence.

And in the vacuum that has created, all kinds of new investment mechanisms have shown up. So we've got 401ks and we've got IRAs and we've got software like Quicken that "helps you manage your personal finances painlessly," it says here. What it really does is something much more modest than that; it allows you to write checks and to pay the checks very simply. It starts out by doing one regular day-to-day activity for you extremely easily and extremely well, which is to say, it shows you a picture of a check, allows you to type in whatever it is you're going to type in, and then either prints it out if you'd like to do that or pays it for you electronically.

So for the first month you use it, it takes about the same amount of time that it normally takes you to pay the bills. The second month it's sort of vaguely magical because you only have to hit the first key of a letter and it automatically fills in whichever payment you made last time that begins with that sequence of letters and so you can do it in a great deal less time and it's keeping track of your budget.

But the important thing is that, Intuit, the company that makes Quicken, designed it to do a whole bunch of other things besides that. It can manage your household budget, and it can manage all your loans, and it can help you to determine when and how you should be saving to pay for kids going to college, or do form inventory or manage any portfolios that you might have or do in future years, one-button income-tax filing, and so forth. This is the promise of Quicken from Intuit. It's a promise that caused Billy Bucks to make them an offer of more than $2 billion to buy a company that only has $150 million in annual sales.

That's an enormous margin, ladies and gentlemen, and there's something weird going on there. The weird thing is that Billy Bucks understands the business about what we call *neo-biological systems*. You've already had a citation in the conference on neo-biological systems. That was when Tom Peters talked about an extraordinary book, Kevin Kelly's *Out of Control*. I'm here to tell you that the *Out of Control* lessons are already going on. This is a list of things that are out of control. They're systems with no centers. The Internet, personal financial management systems, Intel, Windows, movies and videos, licensing of properties—whether it's Michael Jordan or Woody Woodpecker or Casper the Friendly Ghost, e-mail and fax systems, 800 number and 900 number services, and biotech research are all precursors. These are examples of neo-biological systems.

And I think it's important to take just a few moments to explain to you how these things work because they have an incredible new role in the way that they shape people's attitudes and preferences and choices in open markets. And to do that I'm going to borrow from Kevin Kelly a lesson that is in the first seventy-five pages of his book. So even if you don't read the rest of the book, trust me, I'll give you the coolest part right now.

He talks about something called *hive mind*. Which are systems with no center, and the illustration that he uses comes from an example in the natural world: bees. You might not have known this before, but bees don't have staff meetings Monday morning at 9:00 A.M. They don't just say, "Hey, you know it's getting a little crowded in here. Let's build the west side of the hive out a little bit." They just don't do that. They make judgments collectively but without any central control. And the most interesting thing happens when the bees decide that they want to move to a different place because they're running out of room. They do this in the most fascinating way. I never knew this before—you guys proboably all know—but I'll describe it for you.

What happens is that the drone bees go out and start alternative territories. They go hundreds of yards in all directions, and they check out locations for the new suburbs that they might want to move to. And they come back and they try to express their excitement. Now how do they do that? They do it by dancing. The bees dance. So the bees are coming back and you have some bee doing this dance. But another bee is really getting into it, and the rest of the bees are standing back there checking this out. So they basically narrow it down to three or four places, and they send more of these scout bees out to those places. They go check it out, and they come back and there's a totally, really cool dance competition. I'm mean we're talking John Travolta all over again. And there's this gigantic effort to try to figure it out. "He's got good legs but I don't know. He's not as excited as that guy over there."

Eventually what happens is they make a judgment about the best dancer, and then the most astonishing thing happens. There is a bug out, literally, and they all go zipping off to the new territory to build a new hive, and that's how it works. No central control, a lot of experimentation, and a lot of ability to express patterns of preference. And, ladies and gentlemen, all parts of modern life are beginning to behave this way. This is the new way in which people make collective judgments. It's really quite fascinating.

There are also network economies. Look at sports, and you'll be able to get a sense of the interconnectedness of weird things. Madison Square Garden is there beside Nike and Nike Town and Sega Sports and Microsoft Baseball. Michael Jordan gets his own node all to himself. This is what people mean by "network economies." These things are interdependent.

In the same way, video games are terribly dependent on one another. Disney plays a role here, 3DO, Sega, Sony with its new playstation. Blockbuster is playing a big role, virtuality, and Apple's new Pippin and things like that. They're all examples of things that come together and influence one another in strange ways.

In music you see other strange kinds of emerging networks, whether it's CDs by mail, Ticketmaster and Time Warner, and the way they used the

Lion King to sell *Pochahontas*, and MTV and Tower Records and Tapes. All these things are interrelated. I think I can put it into sharper view for you by just talking about the O. J. phenomenon. The O. J. phenomenon is a perfect example of a hive-mind networked economy that just exploded. There are a *bazillion* ways you can find out about O. J. All kinds of news shows and the Internet and faxes and other things like that, but in the bouillabaisse of pictures you hopefully get a sense that it's bigger than the *National Enquirer*. Everybody from the *New Yorker* to the *New York Times* to Internet groups giving you full transcripts and everything else you might want to have. This is an example of how something starts someplace, achieves critical mass, and explodes in many different forms and in lots of different ways.

Which brings us to the final topic that I told you I would speak about, and that is to try to find something that connects the world of design to the world of business that's hungry now for breakthroughs. More things are happening than ever before, ladies and gentlemen. We're changing products, services, and information systems at an unprecedented rate; and I believe that there is a role for thoughtful designers, the designers that aren't just trying to use whatever the prevailing tools are for producing more newsletters. And that role for designers is increasingly and always to humanize the world. This is what designers do uniquely, and this is desperately needed. This is something that will endure as a contribution that designers can make.

But how designers should do that, is a great question. For certainly small projects—an individual business card, the identity package for a restaurant, the posters that are on our seats—are wonderful projects that, if I were a designer and I were able to do them, I would dearly love to spend my life doing. But for many of us, what we will discover—just like the weavers in the late 1700s—is that we're going to have to go through a certain sense of repurposing of our lives. We're going to have to find a deeper level of ways that we can humanize. The good news is that this is not an unknown science.

Margaret Mead was a genius and a pioneer in helping people to learn about and from other people. What she did is she learned to deeply observe the way people work in civilizations of diverse types. And she's

125

got some important tools with her. She's got a pad of paper. She's got a
pencil. And she's got a guy there who speaks the local language. Those
were vital tools to Margaret Mead. And she was able to weave her
particular genius and invent, largely personally, an entirely new kind of
field with those simple tools. She learned how to observe people in
natural settings and to understand the structure of their behavior and to
come up with some insights.

Now this is happening in the design world in various places too. But
already we're seeing the way it's gotten slightly out of control, and
people are already bragging about it more than they're really doing it.
At Doblin Group we use our own methods, something beyond Margaret
Mead's simple set of tools. We're using disposable cameras, digital tape
recorders, lots of video technology, beepers and cell phones and all
kinds of other things.

Sometimes we do really extraordinary pieces of research with very
ordinary tools. We wanted to understand the future of learning and the
technologies that might affect it. We had every third grader and every
fifth grader in two private schools in Chicago take pictures of themselves doing homework at night. And in one week we had three thousand different data points we could use to study environment. And
when we were studying a quick service restaurant chain, which will
remain nameless, we took over six thousand hours of digital videotape
of what goes on with customers and crew members in their facilities.
Lots of companies are out there doing this.

Not a lot of companies are doing what we do in our media lab. When
Ilya Prokopoff draws me an analysis of a case, he uses not only the
digital videotapes, but software that we had to write ourselves. He's got
video printout technology, so he can do a videograph of anything
interesting. And he's got a telestrator, so when teams of social scientists
are sitting in this room with designers they can talk to one another
about interesting things that they observe. And that's how we discover
ways that we can make changes.

Just to put it into perspective for you, at Doblin Group we've had to
invest something like $4.5–5 million in technologies to support the

forty of us who are doing this kind of work. So it's a pretty extraordinary amount of team investments, Michael Schrage, and it does create interesting shared spaces. It also reliably creates breakthroughs.

A furniture system from Steelcase was designed using these kinds of methods. This is a very successful, relatively new line for Steelcase that was all based around their deeper understanding of how people work in teams these days. Both the pacific office space, which was designed around this question of how small a space can you make wonderful; and the shared spaces of the wavy form tables that allow people to get together with their work teams in effective ways were a tremendous success story within Steelcase.

Given that I'm running a bit overtime, I don't want to go through this case too rapidly. I was just going to tell you about a gas station innovation that we did for Amoco over many years. I'd particularly like to credit Peter Lawrence in the audience. One of the reasons I can talk about this case is because he has managed to prevail on Northwestern to make this a business school case and gotten Amoco to agree to talk about it.

What we did is, we systematically reinvented gas stations in order to make them lower-cost, more effective, and particularly more delightful for customers. There were over 125 innovations in these gas stations. Many, many patents, lots of new ways to do payment systems. They're easier to do at the pump, particularly new nozzles. Nozzles were designed twenty years ago, when it was illegal and inconceivable that an individual would pump his or her own gas. Now 78 percent of us, and climbing, pump our own gas. So we designed it like a piece of Sony consumer-electronic equipment with the help of Fitch—people who are in the audience—and created one that had electronic displays, so you could tell automatically how much gas you had pumped and how much money you had spent.

We created graphics with the help of Pentagram and Michael Bierut, and made hand-washing stations in the middle to allow you to get that offensive smell off your hands, so you don't nod off as you're pulling out of the gas station. And we redesigned the interior of the gas station and store again with Fitch's help. And in this case, we were able to make the environment much more appealing and to make it easier to read so that you could tell quickly where the different zones were. We also did a variety of things to make it carry better brands. You wouldn't think it would take lots and lots of years of research to figure out that people don't want to drink Amoco brand of coffee. They'd rather drink Starbucks, thank you very much.

We also designed it along with crew people. Every single person on our team worked in gas stations. It's a requirement. When we try to change the world a little bit, we want to at least know what indignities people are suffering at present. So we all worked in gas stations. It's a horrible job. You get paid minimum wage. You're scared to death somebody's going to shoot you, or that you're going to make a mistake in the six or seven or eight transactions you're trying to monitor simultaneously, or that that guy out there at the pump who is in an extremely big hurry will pump the gas into his car, hop in, and drive off down the road with the nozzle still firmly enlodged in the tank, ripping off the hose and spraying a fairly nasty substance—that's quite explosive—around the general vicinity. When the gas station blows up, people really dislike that. So we had to design around that.

But, ladies and gentlemen, let me tell you this: When people design things, they spend some money for a time. They dig deep in their pockets, and give it to you as designers to invent something. On some happy day it crosses the zero line of dollars and it starts to produce positive revenues. And that's the way the world of business works.

What we suggest that designers need to do is learn how to shorten that time for development even though they do it in a deeper way. That's rapid prototyping, which you heard a bit about already. But at the same time, we suggest that designers need to find the deep structure of human aids, something beyond what you saw in the Ameritech commercials, so you can create things that go well beyond what people can articulate in focus groups. And then to change it so systematically, as you saw briefly in the gas station case, that you can protect the innovation and others cannot catch up quite so quickly. Make it systematic and hard for others to top you. That allows you to have faster time to market and a great deal more area under the curve, so you're getting positive return on investment.

If there's a simple formula for success that I offer up to you today, it's the simple notion that you take some part of everyday life and you simply make it better. You figure out a way, like Quicken did, to make it better and better and to grow over time. Then you make it do the things that people are not yet able to articulate that they want. To be able to be ahead of their needs. And finally, you learn how to extract value out of that in the many new ways in which the world is geared up to generate value.

In our research at Doblin Group we spend a lot of time trying to figure out what the basic changes are that are going on in human's lives. Leisure, retail, travel, home and family, money, work, education, and health; these are eight areas of human life that are in permanent transformation. Designers can make these much more wonderful and much more delightful in enduring ways. They cluster into four simple patterns that are about the way we live, the way we work, the way we learn, and the way we play. Those things are all being permanently changed. They desperately need the humanization that designers can bring and the warm and delightful way that designers can make artifacts that people really love and need in the world.

Designing with the Enemy: Creative Abrasion
Dorothy Leonard-Barton

I've entitled this talk, "Designing with the Enemy," as a take-off, of course, on *Sleeping with the Enemy*, but what I'm interested in is how we can get people to work together on designing new products. Larry just gave you a great context for what I want to talk about: teams that work on new product development and bring some of the innovation he mentioned to life. Now, who's the enemy? Well, the enemy is "them." Now you all know who "them" is, right? Because we're "us" and then there's "them." And if you think back to your last project, "them" could have been the customers, could have been the business group, could have been another set of designers, could be almost anyone; but in a lot of the projects I look at, there's "us" and there's "them." The trouble is getting "us" and "them" together, and that's why I'm calling it, "Designing with the Enemy." As you see from the slide in front of you, there are sharks in the water down below; the problem is that when you're in a tug of war and the rope breaks, you're likely to have problems.

We've also heard from Tom Peters and from just about everybody about the new rules of competition in business. Business people, in fact, feel as if they're balancing all these different balls pictured in my slide: speed, quality, low-cost service, and continuous innovation.

Tom Peters mentioned that Microsoft on a good day is more valuable than the capital market value it sends forth because of its intellectual capital, its capability to invent. So we know that there's a lot of value in innovation and a lot of value in creativity. People expect that designers will bring creativity to bear on problems. I find as I look around and spend a lot of time in a lot of different businesses—small ones and large ones—that rapid innovation in both new products and new services is, in fact, a core competence. Businesses want innovation.

One of my favorite statements comes from the CEO of a mini-mill. It's kind of counter intuitive to think that you're going to find rapid innovation and creativity in a steel mill, but you do. In fact this little mini-mill in Midlothian, Texas, is a very creative, very innovative, learning company. The reason is partially the leadership. I'll quote the CEO, Gordon Forward: "To stand still is to fall behind." Gordon came from "Big Steel"; he said he went to mini-mills because he found that in Big Steel it was like Forest Lawn. He said, "Good ideas were dying there all the time." So he went into his own business in mini-mills, and he's created a very capable, smart company that I'm going to talk a little bit about.

You've heard people throw these buzzwords around like *core capability*, and they are buzzwords; they come in and out of fashion just as black, matte finishes for certain electronics come into favor and go out. The reason I think that companies are looking so hard to find out what their core capabilities are is that they understand that there's got to be something they can build on that will last them and give them a sustained advantage, and they're trying to identify that. And in some companies a core capability is, in fact, the ability to innovate rapidly.

The problem is that the flip side of a core capability is a core rigidity. And they look much the same. A core rigidity is built up over time, as a core capability is; it's not readily imitated or transferred as a core capability isn't; it's based on shared values as is a core capability; but it no longer, if it ever did, gives the organization a sustained advantage. You find companies around the world who have these deep competencies, but they have become core rigidities rather than core capabilities. And that's why they need the transformations that Larry was talking about. They have deep skills but they aren't the *right* skills.

So how is a core capability built? Now the reason I want to focus on this a little bit is that there are some activities that go on in companies, day by day, that create core capabilities and core rigidities. I have found four activities that are managed *differently* in companies that are innovative and companies that are not innovative: (1) problem-solving, (2) integrating knowledge, (3) experimentation, and (4) importing knowledge. All companies have these activities; how they are managed discriminates innovative companies from non-innovative companies.

So for instance, creative companies have a mechanism for involving everyone in creative problem-solving. Nissan Design, which is one of my favorite companies because it's headed up by Jerry Hirschberg—and I find Jerry Hirschberg a fascinating guy—was started about seventeen years ago. Jerry Hirschberg decided that he was going to change things in terms of problem-solving from what he had learned at General Motors. At GM, as you may recall, there was a clay model that was kept under wraps and nobody was allowed to see it. Well, at Nissan Design they do the opposite; they take the model out into the courtyard and anyone can come and comment on it. As Jerry says, "Everybody has fingers so they can figure out whether they can open the door or not."

Well, one time the designers had come up with a brand new design—they were pretty happy with it—they got the clay model, and they got it out there in the courtyard. Everybody was gathered around going "Ooh" and "Aah," talking over the prototype. And a secretary named Kathy Woo came out with her cup of tea. She came out a little late, and she looked at it, and she said, "That's kind of stupid-looking." And Jerry said their whole team looked at each other and said, "You know, we've just been kidding ourselves. This is a stupid-looking design." And they went back in, and they recreated their design and worked on it some more. So they got everyone involved in the process of problem-solving. And that's a characteristic that I find in innovative companies.

You find the same thing in Chaparral Steel; there are just no boundaries as to who's involved in the creative problem solving.

Integration of knowledge is a second activity. Integrating across knowledge specializations is critical. In the integrated companies there's a real effort to integrate across disciplines. I'll talk more about that in a few minutes. But for instance at Chaparral Steel, this little steel company, there's no research and development department that's separate from production. Now in most companies, if you talk about experimentation in production, you cause, if not worse things to happen, at least the hairs on the backs of the necks of the operations folk to go up, because their business is to get things out the door. But at Chaparral Steel they have decided that separating experimentation from production would cut off knowledge to one side or another, so they integrate.

Experimentation is the third activity. Larry just mentioned a few minutes ago that prototyping and experimenting is very important. In an innovative company you really create an atmosphere in which you can *fail forward*. There's a difference between stupid failure and intelligent failure. One of my favorite management gurus is John Cleese, who is maybe better known to you as an actor in movies such as *A Fish Called Wanda*. He gives the following example of an intelligent failure versus a stupid failure. He said, "There are true copper-bottomed mistakes such as wearing a black bra under a white blouse, or to take a more masculine example, starting a land war in Asia."

Intelligent failures are somewhat different, and companies that know the difference and can allow people to make mistakes and move forward are the ones that are very innovative. For instance, let's go back to Chaparral Steel. Chaparral Steel is not a big company; they can't spend a lot of money on experimentation. There's a guy named Dave Fourier who was one of the managers for one of their big mills. He introduced a machine to cut off the ends of the steel bars called a *magnetic arc saw*. He brought it into the mill, and he had spent a lot of money; it was a big investment. Everybody regarded it as a big risk, and it never worked.

It had one small problem: If you got within about twenty feet of it, it would take anything that was made out of metal right off of you and turn it into a projectile. You had all of these pins and watches, hearing aids, and all sorts of things flying through the air. So it didn't work. Now what happens to a guy like that in an innovative company? He didn't get fired. That's number one. He got promoted. He made an intelligent failure, and he got promoted because he was willing to take the risk. Everybody knew he was taking a risk—and that was acceptable.

Importing knowledge is the fourth activity. Creative managers import knowledge into a company through nontraditional channels. A lot of knowledge about new product development comes into the marketing research group, which takes a lot of information and translates it into data and tells you that if you are in a certain neighborhood and you want to sell to that neighborhood you've got to keep in mind that the people who live there have 2.8 children and a collie dog, which may or may not be helpful to you.

One useful set of tools I call *empathic design*, bringing in ideas for new products and services through nontraditional means—including anthropological expeditions. These forays out into users' or potential customers' own environments are very helpful, but a lot of companies don't do them. You know how to conduct such expeditions. Designers know how to do empathic design. What does empathic design mean? It means having a deep empathy for a potential user or a user environment and being able to go out and *observe*, not ask, but observe behavior.

So, how do people observe behavior? Larry just mentioned the personal financial software program, Quicken. How many of you have Quicken, by the way? And it's a great program, right? The company that makes it, Intuit, is worth a lot of money. But how did they get so good? Well, they had all sorts of ways of following people around. One of the things that Scott Cook, who heads up Intuit, said was: "What we're in competition with when we create Quicken is the pencil. That's my competition, and what we build had better be as easy to use."

So one of the things that they did that I think is just fascinating is, they had a follow-home program. Anybody been followed home by somebody from Intuit? Well, I can't imagine exactly how it happens. You go in. You pay your money. You buy your Intuit software packet; and some little guy pops up from behind the counter and says, "Can I follow you home?" And evidently enough people said yes, that they actually could follow people home and watch them unwrap the software package, put the disk in the computer and then get stymied because they didn't have enough room on their computer, or whatever. They watched people in real environments doing real things. They *watched*. They didn't ask them, "Why did you do that?" They asked if they could watch them for a while.

Or take Colgate. If I had time, I'd show you some videos that Colgate had taken in people's basements. Now these are videos that the people took themselves. And they're of, largely, women—this will not surprise the female half of the audience—in the basement doing laundry. And what people do to their laundry would surprise all of you. There is, for instance, one woman who is describing her recipe for doing her curtains and how she starts with a cup of dishwashing soap. And then she proceeds to go through a whole bunch of stuff, including baking soda and bleach, that she has to put in to get her curtains white. Why would you follow people into their homes like that? Because you find out people are using your products in ways you never thought of. That gives you the ideas. Why did the woman want the dishwasher soap in her recipe for curtain cleaning? Because it had a better cleansing agent and was a better bleach, etc.

So there are companies using these empathic design techniques, and I would submit to you that just about everything I've learned from them comes from anthropology and design. So importing knowledge in non-traditional ways is a fourth activity that distinguishes companies that innovate well from those who don't.

Now let's consider the flip side of core capabilities: core rigidities. Core rigidities limit the way that managers solve problems. There's a nice story that I think illustrates this pretty well. Towards the end of World War II, when people in France still needed to protect their coast but were running out of artillery, they decided to use some old artillery left over from World War I that had been in the cavalry. They brought these guns, hoisted them up, and set them seaward, facing outward. The gun crews had been trained by older men who had used the guns in World War I. They had one problem: The crews were slow. The army officers watched the crews that would ignite the artillery, and there was something odd going on. The men would come to attention for two seconds. Nobody could figure out why. So they took some films, and they showed them in slow motion to a cavalry colonel who had retired. They said, "What's going on here? Explain this to me. Why are these guys coming to attention for a couple of seconds when it's destroying the whole rhythm of loading the guns and firing them?" And he said, "I don't know. Let me look at it again." The second time he looked through it he said, "Now I've got it. They're holding the horses." When the guns fired, the horses would start to bolt. You'd have to hold them.

Now, of course, there were no horses there any more. The men were following a very outmoded routine that they had been trained to do. But how many companies have you been in where somebody's still holding a horse? I don't know about your organization, but in mine there are still plenty of people standing around holding the horses. So the trouble is, we're trying to solve problems with outdated rules, and that's where innovation is needed.

The other activities are similarly dysfunctional in non-innovative companies: Specialization and the inability of people to integrate knowledge into the course of day-by-day activities causes the core rigidities that I was speaking of. Limited experimentation, the inability to experiment, the inability to let people fail forward. I was interested to hear from someone who used to work at Sears—and Sears has seen better days, you know—that there were bulletins that people put out to tell people at Sears how to react to things. And my informant said, "Heaven help us if there was a situation for which there was no bulletin. Then that meant it was new. And what do we do about that?" We see companies using the old rules, unable to specialize, not willing to experiment, and finally, they screen out knowledge. They're not willing to try some of the non-traditional techniques for importing knowledge that we've been talking about.

Tom Peters mentioned that innovation occurs at the intersection of planes of thought that haven't been connected before. Most of you know that. That's Arthur Koestler's definition of innovation: "the intersection of hitherto unconnected planes of thought." And that's either "Ha ha!" or "Ah ha!" And it's "Ha ha!" obviously, if you're holding the horses or something, and it's "Ah ha!" if it's a new idea that comes because you've got planes of thought that haven't been connected before.

What I'm going to talk about then is: How do we get there? How do we have that intersection of hitherto unconnected planes of thought? Well, it occurs at the intersection of different problem-solving styles, and that's what enhances the innovation and innovative capability of organizations. What we're looking for is creative abrasion, because what we want is abrasion that produces light not heat. That's why we call it *creative* abrasion. There are lots of different intersections where we might find that creativity. There are different cultures, different disciplines, and different preferred thinking styles. And I want to give you some examples of all three of those in a minute.

First let me clarify that when we talk about creative abrasion we're not talking about when your teenage son comes home and tells you that he has creatively rearranged the chassis of your car, and it now no longer drives normally but sort of goes down the highway sideways because he hit a pole. That's abrasion alright; that's not *creative* abrasion. Creative abrasion isn't personal, and it's not merely conflict, and it requires really conscious management. It requires the people who are working on new product development to be conscious managers of it.

So what is it? Well, let's look a moment at different cultures. Again, let me talk about Jerry Hirschberg. He's working for Nissan, so from the very beginning he knew he had these different cultures with which he's working. And it turns out that that's an opportunity for a lot of abrasion. The challenge that Jerry had was to make it creative. Why is it that we have this abrasion at the cultural interfaces? Well, you can see lots of examples of misinterpretation of speech or something, but there are different examples.

Now probably you know, but I didn't know, that you can design things for smell. And the Infinity J30 had to have a designed smell. How do you want your new car to smell? You want it to smell like leather, mostly, right? So what kind of leather? Ah, now I've got you. What kind of leather? Well, it turns out that Americans have pretty ethnocentric noses, because what Nissan Design did is they put about ninety different kinds of leather in little boxes with holes in the top—the kind of boxes that your kids bring bugs home in—and they stuck them under people's noses, and they said, "What kind of leather do you prefer?" The Americans' noses agreed on three kinds of leather, and they were all American leather. Now if you ask a Japanese person, "How do you want your car to smell?" they look at you incredulously: "Smell? I don't want my car to smell!" They're certainly too polite to say this, but they certainly do not want it to smell like a dead cow. So there's a real difference in how people want things to smell and how people react to things.

Another example of abrasion among cultures is when Jerry first took the design of the Infinity J30 to Japan. He thought they were really going to have to do battle because the American designers were doing what they called "challenging the tyranny of the wedge." At the time, all the cars were high in back and low in front, whereas the Infinity J30 was balanced, low back and low front. And they thought that was going to be a problem. The Japanese did not have any problem with that at all. What they had a problem with was the front of the car.

The front of the car the U.S. designers had designed shows the front grill with a little bit of a down-turn, and they had discovered a new kind of light fixture that allowed the lights to be lowered in height. Their colleagues in Japan looked at this front of the car and said, "We don't like the face of this car. The mouth is downturned and the eyes are (in their words) squinty." And Jerry and his colleagues were puzzled. "What's this about the 'face' of a car?" And they began to realize that to their Japanese colleagues the face of car was the front of the car, the presentation of the car, the entrance to the design, the introduction to the car.

And they began to notice things such as when they went to the equivalent of a showroom in Japan, that people went around to the front of a car first to see *who the car was*. In the U.S., how do you first approach a new car in a showroom? How did you draw a car when you were seven? You go to the *side* of the car. You get in the car. You may kick the tires, and you go around to the front of the car last. So the U.S. designers learned a lot through the cultural abrasion they experienced about the opportunity for *creative* abrasion.

I could give you other examples, but let me just mention Hewlett Packard Singapore, which now has the worldwide mandate for the portable computer printer. It's the first time that H.P. Singapore has had a worldwide mandate for a product. How did they get it? They got it through a combination of design done by an industrial designer from Idaho who was living over there, Jim Girard; their own ability at manufacturing very, very well; and some Japanese software, which they put all together. And there was a lot of abrasion among team members in order to understand each other, but it was *creative* abrasion because they made it so. A second opportunity for creative abrasion comes from different disciplines. We've talked a little bit about that already. What about manufacturing? Manufacturing is sort of a dirty word, I think, in design groups. It's somewhat like marketing in manufacturing. You sort of hesitate to admit that you actually consort with "them"—people who are in the other discipline. But manufacturing has come a long way in the last few years. It used to be that manufacturing was a very passive service organization. But now they're much more involved in design, and there's a lot of abrasion involved in that. Is it creative abrasion? It depends on how it's handled.

I remember walking into a room when some designers were discussing a bottle top with manufacturing. This was a retail product. The designers really wanted the top of this bottle to be square. And they had discovered from manufacturing that it was going to cost them a tremendous amount because it would require all sorts of retooling. And very hesitantly the manufacturing guy said, "You know though, if you kind of cut the corners so you have more of an octagonal top, you could do it, because we have this machine that we could modify such and such and such." The designers said, "That's great, that's OK. That's still maintaining the important elements of what we want." And so there was a meeting of the minds that really allowed for some creativity.

A third opportunity for creative abrasion is occasioned by different cognitive preferences, different thinking styles. This afternoon if you go hear Rich Gold from Xerox PARC, you're going to hear about a program called PAIRS in which they put an artist together with a researcher. I'll leave to him the explanation of why and what they do. But it's an example of creating an intersection at different thinking styles, at the edges of different styles.

Another example is computer scientists using anthropology to design cyberspace. Who can you go to for help, to help you think through what it might be like to create spaces where people can meet in cyberspace? At Xerox PARC they've hired anthropologists to help them think about how people meet, why people meet, what they do when they meet, what the rules of engagement are, so to speak. And what they discovered is that while the anthropologists do not create new products, per se, they do add that element of understanding from a different plane of thought that enables a spark of innovation to occur.

Another interesting organization is Interval Research. Paul Allen, the other founder of Microsoft, started up Interval Research as a think tank, giving it $100 million for ten years, to think about media in the future. One of the things they do at Interval Research in order to get the disciplinary intersection, is bring people from all sorts of disciplines in to temporarily mix up the group. The director, David Liddle, describes them as "the herbs in the dish." They're not the main dish, but they're the herbs in the dish that make the product very special.

How many of you in the audience have ever had one of these psychological tests that are given mostly to sophomores in college? "Myers-Briggs" or whatever? In dozens of these tests, at least four distinct thinking styles show up. The labels given the four differ depending upon which test you take, but basically they look like this. We have four quadrants. In the upper left hand corner we have rational, technical, and quantitative; that's where we do a lot of our schooling. Certainly in the business schools we do a lot of training in that. In the upper right-hand corner we have the visual, conceptual, opportunistic, and entrepreneurial. In the lower right-hand corner: emotive, emotion, expressive, and interpersonal. Lower left-hand corner: organized, sequential, and procedural.

It isn't that we all have only one style. We all have all four. But when we are asked to solve a problem we tend to have a *preference* for how we approach that problem. And, by the time we get through the whole project, we actually need all four. So we need people who are visual, conceptual, and can think intuitively. We also need people who will think analytically as in the upper left-hand quadrant, because they approach the problem differently. And we need people who will implement, as in that lower left-hand quadrant. We need people who will seek out other people to work with (the lower right-hand quadrant).

What happens in a lot of new product meetings is that you see people who are operating in one preferred style trying to communicate with people who are operating in another preferred style and there's a real collision that does not result in creative abrasion, but only in abrasion. Many managers experience this, because they must bring together people who prefer to solve problems from very different quadrants.

Tom Corddry, manager of family reference multimedia at Microsoft, finds very different mental processes essential to produce a product that combines computation and database management with user-interfaces capable of engaging, pleasing, and informing the user. He also needs both people who can "think up" and those who can "shoot down": 1) option creation and 2) option reduction. Although people representing these two approaches can be found in any innovative team, two creative groups commonly involved in creating multimedia products typify the extremes: the most analytically inclined among software code developers—the same people who created Microsoft Excel spreadsheets and can "speak" C++—and the artistically oriented designers of the user-interfaces, screens, icons, and background patterns who think in terms of form, line, and color.

The first group is excellent at logical argument, at finishing things, at getting things done efficiently. The analytical developers, as Corddry says, "make decisions like Ninjas," calculating the probabilities of a particular course of action's being correct and immediately acting upon that analysis. At the other extreme of the continuum, the designers tend to be relatively indecisive and to relish ambiguity. If you ask them to choose between A and B, they will often ask: "What about C?" These two groups potentially miscommunicate—a lot. Analytical people who have traditionally added value and been rewarded on the basis of their ability to reason logically and swiftly to a conclusion may initially assume that people who operate at the opposite end of the mental continuum are not smart because they are often inarticulate about their ideas and both reluctant and slow to drive toward resolution. On the other hand, option creators may believe that highly analytical people are so impatient for closure that they spend too little time on the broad issues of scoping out product concept and that they move on quickly to details, maybe settling for suboptimal design.

Corddry humorously describes the superficial differences between the two groups at the extremes: developers work at night, wear rumpled T-shirts, and eat pizza; designers work during the day, wear "found-object earrings," and prepare gourmet meals. These lifestyle differences are merely surface indicators of very different world views. The concept of something as simple as a line drawn on the screen is totally different to

141

the two groups. To very analytical software developers, a line is a mathematical formula expressing the shortest distance between two points. To artistic designers, a line has characteristics—weight, a variable edge, color, density. Whereas developers believe that a line is a line, designers believe that a line can convey messages about time and space that allow a user to infer function. Designers are skilled in the "art of the desireable" and developers in the "art of the possible." Designers may be relatively naive about what kind of computer power is required to to construct and deliver a function. Would a feature require a supercomputer or a personal computer? They don't (initially) know.

As Corddry notes, "In a disposition-driven culture like Microsoft's, it is important to encourage the imaginative impulses of people who tend to resolve issues almost before they think about them and to attract and hold some people who tend to imagine alternatives faster than they resolve them."*

Corddry's product line forces him to struggle with these different preferred thinking styles, to get those groups to work together. He has some ideas about how to manage it, but when you get people who want to add to the options, they are terribly frustrated by the people who seem to want to close down the options and vice versa. And yet, what I've heard from a number of managers who manage these two groups of people is that it's somewhat a matter of timing.

For instance, Jerry Hirschberg will say, "I'm the kind of person who likes to leap off a cliff with a joyous cry and half-way down the cliff I'll holler back up, 'We need a parachute—*now*.'" And he says,

> People, really, in my group who wouldn't do that, who wouldn't leap off the cliff on impulse, used to really be nay-sayers. At least that's the way I perceived them. They were nay-sayers. I'd get an idea and they'd say, "Yah, well, before you do C you've gotta do A and then you've gotta do B." And I just perceived them as wanting to kill the innovation. What I

learned, was that if I give them a little time, that they will come back
and help me with the solution.

So there are ways for these groups to learn how to communicate if they
understand that the abrasion is not personal, that it's just a different
way of approaching the problem, and that it can be managed. But I
think we're at the cutting edge of understanding how to manage this.
Those of you who do understand these sort of preferred thinking styles
are ahead of the game in that you know you will need to use different
languages in talking to such groups. If you can bring that understanding
to the product-development process, you will be ahead of a lot of the
managers.

So I'm suggesting that to help companies be more innovative, which is
perhaps the next career for some of you if we listen to Larry's figures,
one of the things that you want to do is to help them understand how to
hire from alien cultures. Now you know that if you approach someone
who speaks a different language, and you know they speak a different
language, you pay a lot of attention to them. You think: "Ah, that person
speaks Romanian and I don't speak Romanian, so the accent's going to
be different. I'm really going to listen." But if you approach somebody
who is separated from you by discipline or by cognitive style, but
apparently speaks the same language, you may not put as much effort
into understanding the alien culture.

So can we learn to listen to, maybe even understand, alien languages?
Most of you are very gifted visually, but a lot of people don't understand
visual language. But they can begin to. And that's where the prototyping
and the sketching comes in that Michael was talking about yesterday.
Can we manage both for divergence and convergence? We need both in
product development. We need new ideas, but we need to get that
product out the door. We need to be able to manage both of them. And
can we build on global resources, use cultural differences for innovation? We all are becoming much more global, all of us in all of our
companies. And the question is, can we harness that for creativity?

Finally, I just want to leave you with the reminder that when I'm talking
about creative abrasion, I'm not just talking about abrasion. I wouldn't

143

wish you to go back and say that one of the things you heard at the conference was that abrasion's OK and let's all get together and abrade a little bit. It's *creative abrasion* we're talking about, and I hope we can bring the business culture and the design culture together in a creative way to get that intersection.

V. Reframing Design

Design and Business: The War Is Over
Milton Glaser

When I first came to Aspen, the mantra, "Good design is good business," was the guiding assumption of our professional lives. Although it sounded beneficial to business, like all true mantras, it had a secret metaphysical objective: to spiritually transform the listener. We were convinced that once business experienced "beauty" (good design) a transformation would occur. Business would be enlightened and pay us to produce well-made objects for a waiting public. That public would, in turn, be educated into a new awareness. Society would be transformed, and the world would be a better place. This belief can only be looked upon now as an extraordinary combination of innocence and wishful thinking.

After forty years, business now indeed believes that good design is good business. In fact, it believes it so strongly that design has been removed from the hands of designers and put into the hands of the marketing department. In addition, the meaning of the word "good" has suffered an extraordinary redefinition. Among an ever-increasing number of clients, it now only means "what yields profits."

While we might agree that all of life is an attempt to mediate between spiritual and material needs, at this moment in our work the material seems to have swept the spiritual aside. Hardball is now the name of the game, and the rules have changed. This, of course, is nothing new. The struggle between these issues is as old as mankind. Through the years, as the power of official religion declined, the source and receptacle of truth and morality became "the arts," and all those who were involved in them formed a new kind of priesthood. Designers very often perceived of themselves as being part of this alliance against the philistines, whose lack of religiosity had to be opposed in order to produce a better world.

Now this conflict seems to have resurfaced with a vengeance. One might say that what we are experiencing is merely a question of atmosphere, but the atmosphere is the air we breathe, and it has turned decidedly poisonous. Let me use a recent contract I received from a record company to illustrate this change in spirit. The contract reads, in part:

> You acknowledge that we shall own all right, title, and interest in and to the Package and all components thereof, including, but not limited to, the worldwide copyrights in the Package. You acknowledge that the Package constitutes a work specifically ordered by us for use as a contribution to a collective work. You further acknowledge that we shall have the right to use the Package and/or any of the components thereof and reproductions thereof for any and all purposes throughout the universe, in perpetuity, including, but not limited to, album artwork, advertising, promotion, publicity, and merchandising, and that no further money shall be payable to you in connection with any such use. Finally, you acknowledge that we shall have the right to retain possession of the original artwork comprising the Package.

The first thing one notices is the punishing tone. This is not an agreement between colleagues, but the voice of a victor in a recently concluded war. It reinstates the principle of "work-for-hire," a concept that presumes that the client initiates and conceptualizes the work in question, and that the designer merely acts as a supplier to execute it. It destroys the relationship between payment and usage so that, although the work has been created for a specific purpose (and paid for accordingly), the client is free to use it anywhere, and forever, without further payments.

This violates the most fundamental assumptions about compensation of professionals, i.e., that what something is being used for and how frequently it is used is the basis for determining how much should be paid for it. It also claims ownership of the original art, marking the reintroduction of a mean-spirited and unfair doctrine that we all assumed had been legally eliminated. The overall posture, of course, reflects what is seen in the larger culture—a kind of class warfare that occurs when societies lose their sense of common purpose. The collegial sense of being in the same boat, pulling towards a common shore, has been eroded and replaced by the sense that the rowers are below deck and the orders are coming from above.

The Aspen Conference itself was founded in 1951 by Walter Paepcke and Egbert Jacobson, his art director, to promote design as a function of management. It became, for a time, the preeminent symbol of the modern alliance of commerce and culture. They were joined in this adventure, at least spiritually, by such remarkable figures as Joseph Albers, Herbert Bayer, and László Moholy-Nagy, the last of whom was active in Chicago's new Bauhaus, a school committed to the principles of modernism and the reconciliation of art and consumer capitalism. It is not an overstatement to say that design education in America began here. It is important to remember that the Bauhaus was not simply a trade school, but represented nothing less than the "transformation of the whole life and world of inner man," and "the building of a new concept of the world by the 'architects of a new civilization.'" Cultural reform was at the center of Bauhaus thought, as it has been in many art movements. The Arts and Crafts movement in England, as well as the Viennese Secessionists shared this common characteristic. Modernism itself, in its earliest form, was a religious movement that rejected what it perceived to be the unfair dogma of Catholicism, and attempted to restore religion to a meaningful role in people's lives.

In the United States the social impulses that characterized Bauhaus thought began to be transformed by our pragmatic objectives, such as the use of design as a marketing tool and the elevation of style and taste as the moral center of design. The primacy of individual opportunity and capitalistic efficiency replaced much of the mildly socialist impulses of the modern movement. The metaphysical objectives and the ideal of civic responsibility went underground or were swept partially away. The pressures of professional practice and breadwinning left little room for theoretical inquiry into social issues. Nevertheless, the feeling that the arts in general, and design in particular, could improve the human condition persisted and informed the practice.

In the struggle between commerce and culture, commerce has triumphed, and the war is over. It occurred so swiftly that none of us were quite prepared for it, although we all have sensed that all was not well in the world. Anxiety, frustration, humiliation, and despair are the feelings that are revealed when many designers now talk among themselves about their work. These are the feelings of losers, or at least of loss. The two most frequent complaints concern the decline of respect for creative accomplishments and the increasing encroachment of repetitious production activity on available professional time. These are linked complaints that are the inevitable consequence of the change in mythology and status that the field has gone through.

The relationship of graphic design to art and social reform has become largely irrelevant. In short, designers have been transformed from privileged members of an artistic class or priesthood into industrial workers. This analogy partially explains why the first question now asked about a designer by a client is more often not how creative or professionally competent she or he might be, but how much does he or she charge per hour. If someone is tightening screws on a production line, it scarcely matters that he or she might be a brilliant poet. That person still only earns $15 an hour. The same assumption makes it understandable how a person with six weeks of computer training now can become a designer with significant responsibility in a corporation without having any knowledge of color, form, art history, or aesthetics in general. We once thought of these things as essential to a designer's education.

But why now? What brought us to this unhappy circumstance when there is more design interest, and there are more graphic designers, and more schools teaching the subject than at any time in history? Over-population in fact may be one of the problems, particularly when combined with the downsizing that the sense of a contracting economy and the computer has caused. Economic forces and technology have always driven aesthetics, although sometimes the relationship is not obvious. Aldus, the great Venetian printer, discovered that by setting his texts in more condensed italic letters he could reduce the length of his books by 20 percent. That observation enabled him to save an enormous amount in paper costs and sell the same text cheaper than his contemporaries could. He became the most successful printer in Venice, and the italic style of typography became dominant in books for the next hundred years.

In the past, the design process seemed esoteric, highly specialized, full of internal rituals, and hard to understand from the outside. These

149

characteristics are all typical of spiritual or artistic activity, and serve as a means of protection. The computer, with its unprecedented power to change meaning, has made the process transparent and therefore controllable; and as we know, control is the name of the game. The argument about computers within the field has been mostly concerned with whether they are an aid or a hindrance to creativity. These concerns resemble the semiconscious babblings of someone that has just been run over by a truck. The phrase, "It's only a tool," scarcely considers the fact that this tool has totally redefined the practice and recast its values, all within a decade.

Clients can now micro-manage every step of the design process, and production has become the central and most time consuming part of every design office's activity. The overriding values are efficiency and cost control.

The use of the computer encourages a subtle shift of emphasis from the invented form to the assembled one. Imagery is now obtained increasingly from existing files and sources more cheaply than it can be produced by assigning new work. Electronic clip books have become the raw material for a kind of illustration we might call computer surrealism. Magritte is spinning in his grave, and who can blame him?

The computer appears to be an empowering and democratic tool. The operator can achieve results that previously were obtainable only through the long process of study and skill-development. This partially explains its addictive effect on the user. For myself, someone deeply shaped by old value systems, all expressive forms that are easily achieved are suspect. There are many more bad examples of clay modeling than stone carving; the very resistance of the stone makes one approach the act of carving thoughtfully and with sustained energy. This may also be a small and passing issue. History has shown us that technologies develop their own standards.

Within the field, the internal dissonance caused by these issues can be increasingly felt. Some older designers have resisted the changes in style and attitude that are now emerging—in part from the effects of the computer, but also caused by a generational shift and the general atmosphere of nihilistic relativism that marks our time. This resistance is not always harmful since it often prevents meretricious notions from entering the culture too easily. What is most disturbing is the sour and polarizing spirit that surrounds the discourse.

A new generation of critics has emerged, and there has been more critical writing on the subject of design than ever before. To some extent, the arguments remind us of the dramatic increase in communication that occurs shortly before people decide to get a divorce. The fundamental question of what good design is, and how it functions in society, has come into question from many new points of view. By and large, philosophical inquiry has been separated from professional practice, and those interested in criticism have been essentially marginalized and left to ply their wares in academia and specialized journals.

There is something else to consider that may help us understand where we are: the relationship between the victory of entrepreneurial capitalism, the fall of world communism, and the almost universal collapse of liberal ideology. Here we can see the connection between reduced ecological and social programs, the attack on "soft-headed or subversive do-gooders" (like the NEA and public broadcasting), and our own sense of loss. Flush with success and in the midst of its validating triumph around the world, business is in no mood for accommodation. Recent history has proven to business that unyielding toughness pays, and self-inquiry is a form of weakness.

Unfortunately, with the elimination of an external threat those same convictions have been turned inward. Once again the wisest phrase in the language comes to mind: Pogo's immortal words, "We have met the enemy and he is us." The tendency of unconstrained business to produce a sense of unfairness and class warfare has emerged dramatically, and most of us have been affected by it.

We may be facing the most significant design problem in our lives now: how to restore the "good" in good design. Or put another way, how to create a new narrative for our work that restores its moral center, creates a new sense of community, and reestablishes the continuity of generous humanism that is our heritage.

The war is over. It is time to begin again.

Roundtable Discussion: Reframing Design
Moderator: Jivan Tabibian

PANEL: STEWART BRAND, DAVID CARSON, BRAN FERREN, MILTON GLASER,
DAVID GRESHAM, CRAIG HODGETTS, SAMINA QURAESHI, LORRAINE WILD

JIVAN: Thank you, Milton. Can you hear me?

MILTON: I can, indeed. I can see you too. You've put on a few pounds.

JIVAN: I have to say something in defense of Milton; the technology is making him sound like his false teeth fell out. Even though he is hanging on to some old ideas, he's not as old as that. Milton, you should hear yourself talk all the way here.

MILTON: Well, I'm sorry about that.

JIVAN: OK, what Milton said in "The War Is Over" is basically polemical. There are strong words in his text and some rather extreme positions. The word *punishing* is a strong word. *Meretricious* is a strong word. *War* is a strong word. *Victor* is a strong word. Those are words that not only convey some ideas, but provoke some resistance among those who have not shared his same experiences. So what I want to do this afternoon—because I have some very strong opinions about this subject which I will discuss tomorrow—is simply to see if our panel shares the same perspective and the same assumptions. Milton's talk was based on a very large number of assumptions, or on certain experiences that feed those assumptions. And some of the presentations just this morning would fundamentally contradict the validity of certain statements that Milton made this afternoon.

So, I want to open by asking the panelists to react to certain bits and pieces of this canvas that Milton drew. I want to see if, bit by bit, we can either tear it apart or find what is meaningful in it. I also want to see whether there are shared values among us. It may very well be that the war is not between business and design but between some designers and others, perhaps between one generation of designers and another generation of designers. It may be a false premise, the "business versus design" proposition. It may be more relevant to talk about one understanding of design versus another.

First, who wants to tell me that war is an inappropriate image or metaphor for the relationship between business and design? Now, I know that almost all of you, some more than others—David, David, and maybe Bran and with certain clients, Craig—have a much more direct and ongoing relationship with so-called *business*. (Thank God we don't have to define what business is, design is, or war is.) From what I've heard, it sounds as if not only is there no war, but you guys are fantastic lovers in bed. You are kissing, making love, hugging, learning from each other, teaching each other, and sharing the basic ethos, with mutual learning and mutual contribution.

DAVID G: I think *warfare* is too strong a word. And I won't bother delving into why Milton chooses to call it warfare. But the thing that we've found in working with businesses is that when business only understands design to be some capricious whim about the way that we are going to make something look, or act, or perform, they view it as completely idiosyncratic and mysterious. They have no way to understand it. I don't think good design has ever been about that. I think that's maybe where some inspiration can come from. What gives you the creative spark, what it's always been about in my mind, is collaboration.

The results that I showed at Design Logic or that I showed this morning from Fitch, were those of the huge team that created them, including the client. The thing that I've found is that any breath that I thought I'd wasted on explaining what we were doing to the client—the rationale and the reasons behind it—wasn't wasted. They could understand. I made it tangible to them, particularly in light of the work that we're doing now with a lot of user research.

When you really bring home that this is what the market's asking for, that you've been wearing blinders for two, ten, or fifteen years in thinking this is what your market is or this is what the product you should make is, when you really open them up and show them what people's true expectations are, the way in which people would really use a product (as Dorothy pointed out) then they have a tangible avenue into seeing why you're doing what you're doing. That doesn't mean you pander to the lowest common denominator. We use it as a point of departure for pulling it along, for doing something better than just the expectation. So I think *warfare* is just too strong a word. I don't see it as that.

JIVAN: But from what you're saying, there is not even a skirmish.

DAVID G: There are clients who don't get it. I mean we lose as many clients as we get because we present our work to them, we tell them the process we want to use, and quite frankly, they think we're out to lunch. They think it's simply about styling a box. And we usually don't do work for those people.

LORRAINE: What I wanted to add to the conversation about that contract (and my office has received similar contracts) is that those kinds of contracts may not be declarations of war, but they are symptoms of fear. That kind of wording is coming through the lawyers, and it's about the moment that we're all in when there is a lack of understanding of which formats we're going to be talking about for future use, what kind of distribution, in the context of constantly changing technology. So that you get this blanket language, trying to cover all bases in the known universe. That's what that's about.

On the other hand, I do think that the design profession (if there is such a thing, I wonder) is relatively new and still has to counteract that type of attempt to control with some better sense of what we need and want. If you look at contract law and copyright law in other countries, it's not uniform. There may be other models that we can look at. I don't necessarily see defeat.

JIVAN: In that contract the most interesting part is when it says, even though we buy it for use A, a poster or a book cover, we will have the right to use it for other uses, including the word *merchandising*. So if tomorrow I decide to put it on a T-shirt, I can, since I originally paid for the original art, and, incidentally, I own the original artwork.

Now yesterday afternoon I heard Craig when he was showing some of his recent work. He showed a very interesting, cute, whimsical cookie company. Right? A cookie store that had to fit within a parking space. But he said, had it gotten built and expanded to eight hundred units we would have been very happy, because he said, and I quote him directly, "we would have received royalties."

CRAIG: Absolutely.

JIVAN: But not according to that contract.

CRAIG: That's exactly right.

JIVAN: Explain that.

CRAIG: In fact we came into the crosshairs of your arch enemy, a large corporation. The U.S. Government? No, no, no. Actually one that's closely associated with Seagrams these days. We got into the crosshairs of their legal department. Now, one can say that that is indeed a declaration of war, and this is the way it was related to me: *You are in the crosshairs*. We were shaken, but the idea was that there was some validity in the intellectual property rights to something which begins with a blank sheet of paper, which you're about to proffer to a mega-corporation, and which they are about to then mismanage, and do all the various kinds of configurations that they might.

Now what has puzzled me a lot is that in so many other creative endeavors—music, fiction, dance, genetic engineering, etc.—there is a high degree of regard and respect for the intellectual property rights of the progenitor. Those people can, in fact, benefit from those things over a great deal of time. Even their heirs do, which means families still collect royalties. And those kinds of things are what Milton is complaining about.

JIVAN: David, do you have this experience, or is this just a bunch of old dodos who have it and you don't?

DAVID C: Yeah, it's just old dodos.

156

JIVAN: I know, but I want to make this more leisurely. Is it because the kind of things they do they could claim as intellectual property, and the kind of conception you have of your work is not one for which you can take a "property" approach?

DAVID C: I would back up just a bit by saying I've never received a contract even remotely like the one Milton referred to, except once, and that contract came from Milton. Let me put it in context. I've done this for ten years, and, literally, I've gotten one contract that reminds me of this, and it came regarding a piece of artwork that I commissioned Milton to do. That's my only experience with something like this.

JIVAN: You commissioned him, but he sent you the contract protecting himself from you as client?

DAVID C: It was about six pages of how it could be used and in what way, etc.

JIVAN: Well, so, it's the reverse. It's a way of protecting. Who was trying to protect whom?

DAVID C: I think Milton was trying to protect himself.

JIVAN: Against a client who might otherwise not understand what his rights were? Is that it?

DAVID C: Well, I commissioned him to do work. But I've got to say my first impression was, "Jesus, this is war—what is this?"

JIVAN: What is it in the job culture? The best defense is offense. I guess it is a way of preempting the consequences. But let's be more serious. Craig gave some examples of certain creative types of work where intellectual property rights are taken for granted. And suddenly this is being challenged in some of the design fields.

CRAIG: I'm not sure that's true. I think that you have to look back into history. And you have to look at the realm of ideas that the design world evolved within prior to the advent of mass-communications. We were all fundamentally dealing with static materials. We couldn't send our material over the airwaves. We could never telegraph it. We could never multiply it in quite the ways—prior to the industrial revolution—that contemporary media can. And I think there's been a kind of key change in expectations now.

For instance, when jazz and other music was first recorded suddenly there was the opportunity to press many, many, many records. Someone had to say, "How many times is that going to be played on a jukebox in a honky-tonk in Memphis?" And there was born a guild which protected composers and gave them back, sometimes even through strong-arm tactics, a small bit of money on the popularity of their work—billboard charts, etc. We don't have those mechanisms as designers. We weren't faced with evolving them, and as a result we are, in a certain way, out of the loop. We're just out of the loop.

JIVAN: Yeah, but Craig, the origins of literature are in ancient Greece, and we have had no problem adjusting our ideas to protecting the written word in fiction, or no problem deciding on intellectual property in fiction or essays or the printed word. It's perfectly understood. Writing didn't begin in the modern era; it developed over thousands of years. So I don't know if the origin alone is the question. For ten, twenty, maybe thirty years, there was a period in which the intellectual property for designer's work was either being taken for granted or sort of at last being recognized. And suddenly it's subject to debate. Is this so in your career, Lorraine?

LORRAINE: Well, I'm just thinking, I may be wrong about this, but I believe that the first time the AIGA offered a model contract to its members was around ten or twelve years ago. I also remember that before that, when I was a design student, a kind of underground samizdat trade existed, where a student could somehow get a copy of a famous designer's contract with the amounts blacked out so that you'd at least be able to see how people were writing their contracts. That was the only way you would learn about how to write your own contracts. I'm not sure that this whole discussion should be about contracts, but there has always been contention, and there has never been an agreed-upon standard for designers' intellectual property. I think Craig is exactly right that it has been neglected. Some people have managed to achieve it; others haven't. It hasn't been clear.

JIVAN: OK. The reason we began with the contract was to give a flavor of what Milton's idea is: that there is a hostile, confrontational, mutually non-agreeing ethos and culture between business and the arts, and business and design. That crystallizes the point of maximum friction, but there is a lot more in what he tried to convey to us. One point being, if I heard him right, that in design as he saw it, the good designers—those whose recent ancestry is Aspen and Bauhaus and so forth—had utopian ideas about improving the world, improving our lives and the inner man: That design is motivated by a spiritual dimension; that one uses design in order to create a means for bettering the community, the person, and so forth; that this motivation runs across a different ethos that he calls *commerce* or *business;* and that in that confrontation, one ethos, one priority, one kind of looking at the world, has prevailed and designers now have to abandon the utopian undercurrent and become hired help practicing creativity subject to a profit motive. Is this perception correct, Stewart?

STEWART: I was last at the design conference I think in 1968, when I sort of snuck into the tent.

JIVAN: 1971.

STEWART: 1971. You and I were both there. I remember hearing that design is good—probably changed my life and stuff. And design is clearly good for everybody. That was the message I got. And then, I thought it said design was good for *everybody* to do. That's the message I went away with. And it's in *The Whole Earth Catalogue* and Hackers with Computers and everything else. The hacker update is that they're trying to turn everybody into hackers, and by the way, that was a successful revolution. So my perspective is that it's not that the war is over and we lost; the revolution is over and we won.

I hear the discussion about intellectual property in terms of these contracts. Intellectual property is going to get a lot weirder as the Net takes over everything. This is the ancient past, the guild protection we've been talking about so far. The guild is over. Everybody's doing design. They're doing some good design and some bad design. I think one of the things that designers can do is keep moving the edge of what we mean by good design forward, but it's for everybody. I don't get it. . . .

JIVAN: No, no, no, no. Not so easy. Not so easy. I am willing to fight for the elimination of rights or property rights in a level playing field in which all notion of property is equally challenged. You are not telling me that in twenty years with hackers and the Net and so on that the idea of intellectual property will become almost irrelevant, indefinable, and difficult to presume and defend. As it's happening, you cannot tell me that there are not some people who, in our real world, are greater beneficiaries than others. You are not telling me that just because watching the pipeline made the public good that someone isn't charging for the pipe. You are not telling me that somehow this benefit is as ideally egalitarian as the notion of destroying property rights and lives.

STEWART: I can tell you that, on the Net, where generosity leads, prosperity follows. That's how it works.

JIVAN: Is that a goal, a struggle, or a reality?

STEWART: It's interesting that there was never a slogan. It started as reality. Mosaic, which was created basically for free, and distributed for free by Mark Anderson, has now become Netscape, a very successful business. It's a tool, so it makes it easier for everybody to do design because it's a graphic medium as well as the rest. It's opening up a whole new market for professional designers. And this was done for free. They gave it away. That's how it works on the Net. You give it away then you'll find in time, with the traffic, the business model that emerges.

JIVAN: What type of business model?

STEWART: That's the thing. You don't know in advance. But if you have to wait until you have a business model to do something real on the Net, it'll probably be too late.

JIVAN: So it's self-defining as its reward?

STEWART: You got it. That's the "out-of-control" that Kevin Kelly was talking about. That's what's neo-biological about all of this. You find out as you go what the organization wants to be. The economy of the system's not going to work without compensation, without authentication, without validation, without credit, God knows. But these things are all more liquid now because of the new media, and so if you go in with old forms of how those things have to be accounted for you'll probably lose because there are other people, mainly younger people, who feel generous about it. I don't know. "Let's just try stuff," they say. And then these forms take shape around what they try.

JIVAN: So, business must evolve as well as design.

STEWART: Yeah. Businesses lose in this environment too if they're holding too much to forms that don't work.

JIVAN: So according to you the war may be over, but not only because design is "losing." From what I heard this morning, business is changing. Is this symptomatic of a wider frame in business? Is this just the occasional flower in an otherwise fallow field? Is this a dependable phenomenon? I mean, can we bet on it? Not because I want certainty about the future, far from it. But one still has to live with some ideas about what the scenarios are. It's very good to be excited about the possibilities, but one also understands that terrible actualities do exist. People have to make a daily living.

STEWART: I think Larry Keeley was best this morning on showing some of the curves about how the nature and pace of change has evolved. It is the underlying Moore's Law of the technology getting steadily better in a period of months or years in a really big way—cheaper, faster, smarter, more viable, and so on. As long as that's the case you don't have a stable enough environment for even a present set of rules to last very long. And that's great and that's terrible. That's why it's an interesting time to be around and an interesting time to be thinking seriously about design.

I guess the point there is that the world of the Bauhaus was a fairly static one, and it succeeded and failed to the degree it did in a relatively predictable world. That's not the case now. It's a different world. We're not only on a different planet; we're off-planet. It's not gravity now; it's something else. And that's what makes it really fun. I don't know. Does that speak to your question?

JIVAN: Well, it speaks to my question. There are obviously two impulses when change accelerates. It's like a train that's moving fast. It's probably the train you want to be on but it's tougher to jump on it. It's an exciting train but. . . . I hear this "easy-entry" business, that somehow the new technology allows an easier and cheaper entrance into the field. Is that notion really true? On one level, six weeks of training, $2,000 worth of equipment, a few hours of learning from somebody else, directions back from other people on the Net, all those things makes entry objectively easier. And yet, somehow, there is a sense that if I don't enter now I'm falling behind. So the psychological cost, the opportunity cost, the psychic cost of being left out gets bigger.

STEWART: Well, the platforms keep changing. If you missed this platform, just wait two years. Catch the next one that goes by.

JIVAN: But people who miss one platform usually become less apt to try another platform. What was it that Mark Twain said? "A cat that has sat on a hot stove once will not sit on another one, even a cold one." So what happens is that we have a greater and greater number of people who feel marginalized by that rate of change if they are not able to capitalize on it. This is a real social phenomenon.

STEWART: I would agree if it were the young people who were feeling left out, but that doesn't seem to be the case.

JIVAN: Only those young people who have access don't feel left out.

STEWART: Yeah.

JIVAN: I guess now I understand why the Speaker of the House wants to give a laptop to every child. I imagine that unless you also feed them virtual food they have to have food first and then laptops. Maybe not. I don't know.

SAMINA: Oh dear.

JIVAN: "Oh dear," is right.

SAMINA: I want to just respond to the fact that we live in a time of profound change, to embrace what you have said. And though I have the greatest respect for history, pining for a golden age isn't going to get us anywhere.

JIVAN: Assuming it was golden.

SAMINA: Assuming it was golden. I think that no matter what the era, ideas remain a powerful force. Power will accrue to those who have the intelligence and vision to make themselves heard. So I think that all of the things that you were saying about intellectual property and the way things are changing is what it's about. And these debates are appropriate in government also, because the voice of the people is demanding change. I think that has to be processed and understood, and I think people will only get what they deserve through demonstrating a broad, and not just a technical, intelligence. It's all about the power, the power of the idea. So that's why I just don't think that talk about a war, or having business and design juxtaposed in that way, is appropriate.

CRAIG: Well, you can sympathize with Milton. I think we all can. Because he's mourning the death of a culture. He's really mourning that the culture which actually descended from a kind of aristocratic agenda, of propelling things to a high level of achievement on a kind of patronage basis is gone. It's just gone. It's now a free-for-all. And one of the reasons that in a funny way I can feel sad about that, but I don't feel the shock to my system is because I've lived on the West Coast for so long. Frankly, the big shock of moving to the West Coast was that there was no establishment. There was no culture; there was no established way of doing business. The only model was the fame-industry model.

The fame-industry model is very interesting. I come back to the war thing. When you make a movie, it's like making war. You get together a group of people who have never met one another before to go out and in a very, very limited time they make the movie. And then they all go away. There's no scar tissue left after that. People's personal relationships erode; they go on to the next project and there's not a kind of business formed around that from the creative end. The only business is distribution. That's why every time one of those things is sold, it's the distribution. It's the rights to the old films which are valuable.

JIVAN: David, is this a tradition with the new East versus West mourning an old culture as Craig said?

DAVID C: I would agree with that largely. I think there's certainly more structure traditionally on the East Coast in business relationships than there is on the West Coast. I think that's probably a part of it. I think Milton's essay is largely a mourning of an era and quite possibly a personal one, and I think that's what we're hearing. I think anybody can relate to that and possibly feel some sadness but to yearn for the return of these, as somebody said, global modes of the great old days seems kind of pointless. I think, partly, it's probably the structure, but I wouldn't write it all off to East and West Coast or left and right coast.

LORRAINE: Or even history for that matter. One of the things that really fascinated me in Milton's paper was that he brought up Aldus Manutius. You know he talked about Aldus Manutius in Venice in 1490 or 1510, working to make a more efficient letterform to be able to squeeze the printing in. Wasn't Aldus being extremely entrepreneurial? I mean, what is that if not hardball? Now at the same time, Manutius was a commissioner of manuscripts, an editor, a publisher, and a distributor. If you look way back into history, there are examples, especially in that first hundred years of printing when it was essentially a new media, of people working in a very different way than the kind of specialization that characterized design after industrialism. So, you know, I don't see history as either hopeless or irrelevant.

JIVAN: So we shouldn't get too excited because we aren't the first or the last to break through this transition.

What about the explanation of mourning a personal loss? Because Milton is claiming more than that, that it's not a personal loss only but a real symptom when he talked about "moral centers" and "generous humanism." I mean, are they disappearing like everything else? Or is this irrelevant because all people couch their personal loss in their defense? What's the globalized consensus, Stewart?

STEWART: This is fascinating. I think there's a really genuine political dimension of some depth here. And Mr. Glaser's statement speaks to it. Because I hear in it a Bauhaus-Modernist left. Tom Wolfe called it socialism, a political framework.

JIVAN: *From Bauhaus to Our House*, yes.

STEWART: And this was a world view. It was a political world view that the good of the people would be served in a kind of a-centralized way. It is absolutely the fundamental nature of the old hackers, the creative personal-computer revolution, and now the new revolution of the Net. The political underpinning is libertarian, which is, if you stay on the left-right spectrum, probably more right than left. And so, oddly enough, you will find in this part of what Newt Gingrich is interested in—more sympathy on the Net with Newt Gingrich than Al Gore.

JIVAN: Of all people.

STEWART: And he's basically a Nethead. So I think there is something really fundamental happening here. And again, you look at Kevin Kelly who wrote *Out of Control*. He is a right winger. George Gilder, a right winger. These are conservative people, but they're not predictable conservatives. They're neo-conservatives in the sense that you can't trust them.

JIVAN: Uh, huh. They're following where they think—

STEWART: —evolution wants to go. There's a wonderful new book by Daniel Dennett called *Darwin's Dangerous Idea*. It's actually, to some extent, a political debate about evolution between people like Chomsky and Stephen J. Gould who are sort of on the left, saying, "No, no, evolution can't be that out of control; that's too weird. There's got to be some idea that's driving everything." And on the Net there's not that much of an idea that's driving everything, and so it's a libertarian world. It's a politically different framework. There are probably lots of serious errors in it because of that. But it is the spectrum.

JIVAN: Does the possibility of a value-free evolutionary view, justify that individuals should therefore accept the possibility of a value-free evolution for human affairs? Should we say, "If that's how it is, that's how it is"? Or is there something that, not just in the West but in every tradition, over thousands of years of history, says people have to make personal and public choices, ethical choices. Is *ethics* a taboo word in a libertarian context? Is ethics a leftover?

STEWART: Libertarians like the First Amendment. They think it's unethical to introduce censorship.

JIVAN: Well, is that a good use of the term *ethics*?

STEWART: I don't know.

SAMINA: I wonder what Milton thinks about this?

JIVAN: Milton, what do you think?

MILTON: I feel, once again, victimized by technology. I can understand about one third of what's coming across. I hope that you can understand somewhat more of what I'm saying. I have a relatively simple point, it seems to me. But it does not seem to be sufficiently clearly articulated. I was saying something quite simple. I was saying that business has now, in its control of the global economy and the full height of its power, recognized that its inherent value system, which is essentially capital accumulation and the pursuit of profit, is a worldly view of life. And it no longer has to accommodate itself to any alternative views.

What I'm saying is that those of us who started out with notions of aesthetic beauty, art, religion, spirituality, and whatever else you want to call the impulses that got us into a practice that was involved with the pursuit of beauty, have a somewhat different value system. I'm saying also that there was a time when we could exist somewhat in the cracks and basically be effective in some way in offering a countervailing view, but I'm saying that the cracks have disappeared and that it's very, very difficult to offer to our clients, to the masters of the universe, an alternative to their incredibly successful accomplishments in demonstrating that the pursuit of profit is a valid, and perhaps overriding definition of what business should be doing.

I don't happen to believe the pursuit of profit is the only way to approach life. And my sense is that there was a time when the alternative to this had more vigor and more resources than it seems to have at the moment. And that, I guess, is what I meant when I said, "The war is over." My sense is that profit is going to drive the Internet. It seems to be a democratic or egalitarian opportunity for making everybody their own communication resource and making everybody their own designer, but its real growth will occur when business discovers how to make a profit out of it. At that time you will see a growth, an expansion, and a distortion of the Web that has not occurred yet, simply because they haven't figured it out. Enough for the moment.

STEWART: What's interesting is, in this point he's absolutely in agreement with nearly everyone one sees on the Net. There's fear that somehow Microsoft, or AT&T, or some entity, or the government will take control of it. But there is the existing fact of twenty-five years of generosity and extreme interactivity and of people creating and basically changing their lives and building their commercial lives around the peculiarities of the Internet. So for Microsoft or anyone else to show up this late in the game and say, "Well, that was fun, fellows. But actually we're going to do it this other way now. You're going to have to pay; you're going to have advertising," that it's a commercial game, "Do it our way or don't play." It's too late, it's not going to happen that way. Twenty-five years are already in place. The Internet was designed to route around atomic explosions. It can route around corporations.

JIVAN: Does everybody on the panel share that optimistic view with Stewart, that if it could handle that it can handle this? Or is he being classically optimistic? You work with people who won't let us be that way.

BRAN: No, I don't think that's true. I think that what's exciting about the Net and all of this other stuff at this particular moment—more so than twenty-five years ago when I was first on it and it was basically just people in academia or the military talking to each other because those were the only people allowed to use it—but what's interesting to me is, and I guess I often have a minority opinion, whereas most people appreciate order and things following a set of rules, I find beauty in chaos and things being completely out of control.

My sense is, with the Net and with what people are attempting to do with alternative networks and such, that it's a great chaos lab going on whereby almost anything that you can attempt to do now analytically is, by definition, invalid. It's sort of like if you take two points in dough that's being mixed, the classic chaos problem. After about two folds of the dough there's no mathematics that can predict it.

What I find exciting about what's happening with the Net and with new media is it's just like that. It's a dough that is in a state of chaos. That's what's so much fun about it. Because the fact is anyone, any one designer, can be the equivalent of the butterfly flapping its wings on the other side of the planet that creates the hurricane. That's an exciting time. This technology has enabled that with a clarity that in the design medium hasn't been possible before.

We hear a lot of discussion about how you can now enter the world of design much less expensively. You can enter the world of design with a piece of paper and a pencil. I don't know why someone thinks a $2,000 computer is a cheaper way to start thinking about design and becoming a designer. Now granted, desktop publishing has given us a more efficient and economical way of doing bad design within the corporate environment. That's a contribution, I guess. But at the same time, the issue to me is not *entering* design. It's what you do after you're there that matters. Getting in is only the first part of the problem. Doing something good remains the challenge.

STEWART: So we've heard from corporate America. Now what?

JIVAN: Is that typical of corporate America, that with a pen and pencil you can enter through the doors of business?

BRAN: I've always thought of myself as being typical of corporate America, so from that perspective I'd say yes.

JIVAN: Actually the better question is: Is business typical of corporate America?

BRAN: I think that is an interesting question. To me, corporate America redefines itself based upon opportunity. The company I work for does one thing for a living. It tells stories. And what it does is bring people together who know how to tell stories in a variety of different forms, and it charges money for that. To me, coming from running my own little business which was a design-oriented and technology-oriented business, I guess it's nice to think that it's evil to have a business structure, that the notion is that you make more money than you spend. But I found this convenient as a person running a company.

And nor was the notion that my objective was, in fact, to make more money than I spent something that meant that I wouldn't put the whole company at risk to do something that I believed in and thought was important, or thought was good, or thought was interesting, knowing there was a chance you'd lose your shirt and your house and your car and everything else that goes along with it. I think there will always be people who have passion about things that encourages them to take risks, and I think that's always been the case and always will be the case. I don't think that's an artifact of business or how business changes.

JIVAN: I like it how we admit to talking about complete irreversible fundamental change, and yet we so quickly fall back on phrases like, "It has always been like that, and it will always be like that." On the one hand we talk about chaos and the inevitability of change and, on the other hand, we go back to certain fixed notions of how society and organizations have to operate, seek profit, make more than they do. It's really amazing how in the midst of all the change there are some very fixed ideas. Do you notice that, Bran?

BRAN: Yeah, but I think you're trying to make something profound out of something which is perhaps less so, to me.

JIVAN: Not profound? Contradictions. Some contradictions don't have to be profound. They are based on consequences.

BRAN: I don't find that when one is in a discussion of the trajectory of design or business and how it changes, the fact that I need to take a breath every twelve seconds or so—and that will always be the case, and has always been the case, because it's part of being a human being—is necessarily contradictory. And I think the same thing applies here, where there are a set of human values—call it morals, call it philosophies, call it a bunch of other things—which are subject to change, bending, moving, changing form.

At the same time, the notion that there are, in fact, common values that can help guide you through a time of change, such as that you care about quality or such as that you care about art or other things, that is the sort of thing that I think I can take comfort in occasionally. To me, if you're doing something new that you don't understand—new rules— you don't understand the other people's rules, I think the fact that you can take comfort in your value system to pull you through it is kind of nice.

JIVAN: Oh, more than nice. I'm glad there is that side in which one takes comfort in the sort of things that one makes choices about as being less than subject to daily negotiation and change and chaos. But don't misunderstand me, I'm fully in agreement with that half of it. It's just that I want us to hear ourselves say that we do still have some choices about what we want to remain attached to in a world of—in this *maelstrom* is the word—of constant change. Where does government come in? Is it for change or stability?

SAMINA: I think that the government is for change. Certainly that is what I hear from Congress. Everybody wants change. But at what expense and how? That is the question. How is this new technology, as I was asking Tom Peters the other night, helping to reach people who are making all of these decisions without all of the necessary information? I think that is a very profound question. I think that the intersection of all of this with politics is a very interesting question, something that we haven't touched upon very much. And I'd like to hear from you. What do you think about that?

JIVAN: Not me. They haven't given me that right yet, that's tomorrow. David, do you think government should, in defense of certain values and ideas, stick its nose in? And do you think that if they had better information they would make better decisions? I mean, I think that some of them think they have all of the information they need to make the decisions they're making.

SAMINA: I disagree with that.

JIVAN: They say they know everything they need to know about it.

SAMINA: Well, I think the way that decisions come to us makes me believe that the decisions are based upon information that has been gathered throughout the country, I'm told. But I don't necessarily believe that all of the information is being processed without a vantage point. And there are constituencies and there are realities in the political universe that are very different from corporate America. So it's really two separate situations, and I think that bringing them together is another whole design challenge. I really think that a lot of these decisions could be very much informed by the kinds of things that I have heard from all of my colleagues here. I don't see that as a situation that is accessible, at the moment, at the government level. Do you believe that? I mean, may I ask you?

STEWART: Yeah, the role of government I think is potentially really interesting. You probably don't want government in a realm like this to lead completely. On the other hand, it was government that started the computer revolution, basically, with the Pentagon saying, "Computers are going to be of the essence. We'll probably buy most of them. We'll fund their getting out ahead of everything else." And that worked pretty well. They funded the original ARPANET which became the Internet and got out of their control. And they actually don't worry about that.

You probably want some laws in your society. You don't want government so far behind that they're so irrelevant that it doesn't count anymore. So you want them in the sense of regulation, catching up. We have a very good techy Nethead in the White House in Al Gore, and a very good techy Nethead in the Congress in Newt Gingrich. At the worst, they'll cancel each other out and nothing useful will happen. At the best, they'll actually supplement each other in a useful way. The dialogue will rise to a more relevant level where it might actually be useful. Does that speak to—

SAMINA: Well, what I'm wondering about is: Is the cross-pollination between everything that I've heard at the conference and the private sector being brought to advantage, to give the government the kind of information and knowledge that would benefit the people? After all, it is perhaps an old-fashioned view, but it is my belief that the government exists for the common good, and that some of the actions of the government don't seem to me to be apparently for the common good. We have heard our chairman, Jane Alexander, speak about the National Endowment for the Arts as one small example. But it's the example that I am intimately involved with and, therefore, concerns me greatly. And that is what design can do to convince the government that it can deliver more value for less cost. I mean that's something that would really make a great deal of difference to Mr. Gingrich.

BRAN: I think part of the problem is that government is often an incubator but also an *idiot savant*. It may be the thing that incubates best, but it has no clue that it's doing it. Therefore transitioning something into something else is awkward at best. The Internet was certainly an example of that. There was certainly no plan to create a tool for the people to broaden interpersonal communications. That was not the design brief for the Internet.

SAMINA: You see, I think that's just the point. How could we, as designers, put forward that information so that it makes government relevant? So that all of these wonderful things that the government invented in the first place, like the Internet, can serve it instead of work against it? What about shortwave and radio and all of that that's out there? It's the same sort of information network.

JIVAN: Bran, when you say government as an incubator is an *idiot savant*, do you mean there are less-idiotic *savants* in the incubation business when compared to government?

BRAN: Well, I think that the role of government as an incubator is one that was never self-defined by government or by the people. I think that there are people who think of themselves as being incubators who tend to do it better, because at least they acknowledge that that's the business they're in. I think it would be healthy in fact to occasionally get government to start thinking of itself as an incubator of ideas and such. I don't see that model happening terribly often. I see it happening but usually, again, as an indirect consequence of some other goal.

JIVAN: I was starting to ask David a question, but a word that Samina used can come even closer. Is the notion of *benefit* relevant? I mean she asked, "Who does it benefit?" Is that word relevant in good design?

DAVID C: Well, before you get to good design, which of course is a whole other topic, I think it's a scary time for this to be becoming so prevalent and important—at the same time that we seem to, as Milton says, have taken a very conservative bent. So if, in fact, government's going to be getting more involved in this type of thing, for me it's happening at a very scary point when a lot of the more liberal ideas seem to be on the way out.

I think what concerns me more as we talk about all this great information and our access to it is that in reality there's a large section of the population that doesn't have their own computers and laptops. And this gap is just getting wider and wider; so we have a certain elitism among people who are totally tuned in and think it's wonderful and have access to it and understand it. Meanwhile this other group of society is getting left in this gap that is widening dramatically. I think that's a bigger problem than defining good design.

STEWART: Is there a design solution to that problem or is that strictly a political problem?

DAVID C: I suspect there's a combination. I'd like to think it could be design, but sometimes I think design will save the world right after rock-and-roll does. If it's strictly political, that is a little scary to me, too.

JIVAN: But if it's not strictly, it is somewhat political, right? You see, it's very curious, David's clear ambivalence based on questions of equity, who's left out, egalitarianism, elitism—the fear that if you really listen too much you might end up on the wrong side. All of us are suffering from this. I want to ask you a question. Designers are like anybody else; we think in metaphors. And I'm very aware this week that evolution has become a very powerful idea, that somehow the system is evolutionary. There is change, incubation, eggs being laid, giraffes sitting on the eggs of ostriches, ostriches going and attacking elephants. What about the consequences of the images we use to understand ourselves and society? You see, I, personally, see the convenience of it; but I also see the conceptual as well as ethical consequences of viewing the world of human beings organized as an evolutionary process.

STEWART: Be specific. Say why evolution could be a bad metaphor.

JIVAN: Well, for the same reason that what's great about evolution is that there is no central plan. There is really no creator or package manager, and it is loaded with accidents and with random events and adaptation and so forth. And the outcome in evolution is basically neutral. Nature will not give a damn if all of us desert it tomorrow. Nature won't; we might. We are a subsystem within a system. And we are a subsystem that presumably can make choices. Dinosaurs weren't consulted on adapting their behavior. Nature takes care of all that. So with what image should we participate in tinkering with our own evolution? That's really the question. Because the evolution immediately takes us to the libertarian laissez-faire idea.

BRAN: The whole process of evolution includes our tinkering. It's convenient to think that evolution was a convenient development that got us here, and then it stopped now that we're perfect. But the fact is we are continuing to evolve. Design does not evolve. *We* evolve and design is a mirror of the designers who are the product of the evolution. And the notion that we have the power to kill ourselves off is part of the process of evolution. Because what's left will evolve into something else. It's not as if evolution is a process that you take and you say, "Well, I'll use a little evolution here but now I'll tamper with it." The fact that you have the ability to tamper with it is intrinsic to what evolution is.

JIVAN: But we do that. We do it daily. Daily. Think of it.

CRAIG: But we're at a very interesting crossroads now because I think one of the things that's so exciting about the Net—even though I can't even navigate it; I truly can't—but what is fascinating about it is that we're approaching a moment—of course a hundred years from now people will look back and say "What idiots!"—but we're approaching a moment when we can have chaos. Prior to this we were really in a highly structured situation. If you wanted to set up a manufacturing plant, you had to put it on a river, and you needed to have the water wheel there. If you wanted to talk to somebody, you had to walk to their house, etc., etc. The structures, the kind of lame structures which govern all the things that we do, are finally getting to the point where they're fairly fluid. They're also fairly able to be very responsive in nanoseconds. So that all these transactions are happening at a rate that is basically ungovernable. And that's a very exciting place to be, because that begins to say, "You can't steer the ship; the ship steers itself." We don't have any real choice at a certain point because the structure is so fluid.

JIVAN: The two halves do not go together, Craig. Just because you can't steer the ship, it does not follow that the ship steers itself.

CRAIG: That's evolution. This ship did. The dinosaurs died.

STEWART: There's a reason for that. It's because hindsight is better than foresight. There's actually something to look at. Evolution operates entirely by hindsight. Some of this came out of this six years of research I did on buildings. Why are vernacular buildings so wonderful? Why

175

did that book, *Architecture Without Architects*, sort of blow everybody's mind in the sixties? Well, vernacular architecture is basically evolved design. There is such a thing as evolutionary design which goes incrementally, looking backward at what worked, building that, and dropping what didn't work and so on. This is now up against the very interesting problem of Moore's Law, where things move so quickly you don't have enough stability for evolutionary design to have enough to base its hindsight on. So you may get some pretty strange and heavy failures. Chaos, among other things, is capable of very large excursions that you didn't want just now. And it might take out the Net. It might take out the financial system. There are some dangers in all of this.

I think one of the interesting things to think about as designers is that evolutionary design is often pretty good; and, I claim, healthier than visionary design in many cases. How do you do it in a fast-changing environment? I think scenario planning—as opposed to, say, programming—is one solution to this, where you give yourself the opportunity of anticipatory hindsight. You consider that this world, that world, and the other world might come to pass during the life of the building you're designing. How would it fare in each of those worlds? You look back with a kind of a dream hindsight. And you may get better design out of that. That's my hope of how we may be able to make design that is more comfortable in this fast-changing world.

JIVAN: Milton?

MILTON: Jivan, I guess I'm still struggling with trying to make a few simple points.

JIVAN: They're not welcome, that's why.

MILTON: The other day when I looked at some reproductions of new cave drawings discovered in France, I realized, to some extent, how fundamentally unevolutionary art has been in the sense that those earliest expressions of human consciousness were powerful, meaningful, and as fundamental as anything that's ever happened since. To some degree you can say that while art has changed in its most fundamental way, its ability to affect us has not evolved very much. I found

176

myself as moved by those cave drawings as anything I've ever seen in my life including the art of the High Renaissance or Picasso.

I think one has to be careful about the nature of evolution in terms of establishing what parts of evolution in fact are useful to us, and what parts are a danger. Things don't always evolve positively. Also, the most fundamental things about human life, the nature of affection, loyalty, friendship, etc., do not seem to have evolved very much in terms of human needs for those things.

My apprehension has always been that with a blind acceptance of technological change particularly, more will be lost than we had bargained for. I admit to a Luddite sensibility that suggests to me that technology, because of the fact that its consequences are unknown, has to be approached with great reluctance and skepticism. I do feel that the velocity that we are moving at is extraordinary and out of control. But that is the nature of technology.

The suggestion that designers can do things or designers can control things is a bit of a misunderstanding and precisely the basis for my initial presentation. But the fact of the matter is that designers are becoming, it seems to me, in many areas, increasingly less significant and increasingly less powerful in terms of modifying the course of events. It seems to me that designers are becoming increasingly business-franchised—that it is, in fact, the triumph of business and business' objectives that have basically designed our future.

Precisely the point I was trying to make is that the design of our future now is not in the hands of what we call designers. It is in the hands of the multinationals, corporate institutions with enormous power. And we fundamentally play a very small part in that enterprise. I would like to see that consciousness in some way modified to acknowledge the fact that at this point in history our role has actually become pretty much subservient to the larger business and corporate enterprises.

JIVAN: If you lose the power, you cannot dictate or define the discourse.

STEWART: Everybody knows power is leaving the guild designer. I'll bet

177

you know who he has lost power to; he has lost power to people who do their own design. I designed the cover of my buildings book. I designed every page in it. Not because I'm a hotshot designer, but because of PageMaker 4.0. I should have used Quark, but PageMaker is all right. The production cost less than the publisher would have spent. The readers are getting exactly what I wanted them to see. I put some poor designer out of business. But that's great, I think.

JIVAN: Is that really what he's saying? Is it that simple, Stewart?

STEWART: Well, let's ask Milton. Is it OK if I design my own book? Is that all right?

JIVAN: Milton, you heard it. Stewart is asking if it's OK for him to design his own book.

MILTON: The only question I would ask about your book would be whether it was designed as well as it could have been to communicate effectively to the audience you hoped to reach. My presumption would be that if you were a good designer you would have done that job well. If you were a poor designer, you would have done it poorly. You may be gifted enough to deal with that problem. Unfortunately, there are still enough people who don't know how to deal with it, so there's some work left over.

STEWART: But look at the design problem here: 350 photographs, high levels of text on every spread. This is pretty demanding stuff for me to be trying to do on the phone with some poor art director in New York. You've got to have it close by. I could have done it with somebody sitting right next to me; but in this case, I was able to lay out the text of each chapter, then lay out the pictures that I wanted to be in that chapter so they're evenly spread through it, then I wrote and laid out the legends and the captions and the credits that went with them so it all made sense as a unit.

LORRAINE: I do that every day. And I make legible books that are visually beautiful, too.

STEWART: Of course you do because you're a designer. But there are tens of thousands of designers now instead of just tens.

SAMINA: You should have organized the information and given it to a good designer.

JIVAN: On the other hand, the strength of Stewart's argument is that as long as he can define what the task consists of he is the master of his own universe. I mean, if as long as you define the design parameters of what you did in terms of what you described, how can you lose? But it may very well be that a designer's designer may have seen other dimensions to that challenge. The designer's function is to explore the dimensions within which it must operate and not reduce them. With all due admiration to the work and the book, it is a reassuringly reduction- ist definition of what the design of the book is—effective, efficient, and constrained by its own ambitions.

STEWART: OK, let's go back to collaborative work then. The standard truism in magazines is: Designers won't read; editors can't see. This is a famous problem, but it's changed a little bit. By and large designers still won't read, but a good many editors are learning to see. The shift is partly right there.

LORRAINE: Yeah, I guess this is the very thing. I guess what I'm actually hearing is a logical response to a lot of crummy design. And actually a lot of that crummy design is often foisted off on people in the name of a kind of a personal creative vision that has a sort of mysticism about it. It's what David started out describing, just saying, "Oh, it's my idea, my vision."

If there is a generational dimension to this whole argument, whatever it is we're talking about, part of it is that for those of us on the younger side, we actually know that that kind of mysticism is not going to cut it any more. And it is all so much more interesting, the whole issue of being a skillful collaborator, of being passionately tied to the problem at hand rather than to the creation of one's own persona based on a vague morality, tied to a vague interpretation of modern-ness that isn't even historically true. All of that just doesn't work any more. So there.

If you want to hire a designer, there are lots of really wonderful young ones; there are actually wonderful old ones too.

JIVAN: The time is upon us. I would like to thank you all. My comment is—Milton, you wanted to say something?

MILTON: I just wanted to say something else. We addressed this question of the fact that it is true that you can lay out a book and that everyone can learn Quark and everyone, to a large extent, can learn the methodology for producing layouts. Layouts and typography are not design. The ability to lay out pictures in a magazine or to lay out a book is not the boundary of design. Although I must say that, to a large extent, that has become one of the definitions of what design is.

Look, there is a distinction between playing the fiddle and playing like Heifitz. And you can hear it, and although they both might be playing the same tune, there's a qualitative difference in the response of the audience. You have to measure the distinctions of when it's worthwhile and when it isn't worthwhile. What I'm suggesting is that there is no question about the nature of collaboration in design, and the need to talk through problems in design, and that design is by and large a utilitarian activity that deals with practical objectives and must meet them. I'm talking about something a little different. It has to do with another quality that design potentially has, that of enlarging the listener or enlarging the viewer, so that the experience is not simply functional, but enters into another domain.

Now this part of design may not be very fashionable at the moment, but it is an element of design that is extremely critical and valuable. And to parochialize design to essentially a process of accomplishing simple tasks in the most utilitarian way does not seem to encompass this larger potentiality, even though this larger objective represents a very small percentage of the activity in professional terms. While there's no question that the largest part of the design activity is utilitarian, perhaps the most significant part of design exists outside that definition.

JIVAN: Thank you, Milton. And since you had the first word, we'll let you have the last word.

VI. Quo Vadis

The Twilight of Humanism
Jivan Tabibian

Good morning. I know that some of you were here at 8:30 last night struggling with our teleconference with the Japanese. If it weren't ridiculous it would have been humiliating. It was a struggle, and I kept rationalizing that my grandmother used to say, "Every barber needs an apprenticeship on a bald head." In yesterday's case I thought it was a little more appropriate that my best and closest friend, Milton Glaser, is quite bald. I just felt bad that this apprenticeship had been at his expense. And I just heard something which leads me from a humorous comment on that phenomenon to a slightly more interesting one.

Last night at 11:30—unbeknownst to me because I was enjoying an illicit sin against all the rules and laws of this incredibly health-obsessed community—while I was smoking my cigar, apparently there was a second attempt to communicate with our colleagues in Japan and it was quite successful. This brings to mind the observation that from the earliest period of Greek mythology, the gods were endowed with an uncanny ability to tell the true believers from the fake ones. The gods have always punished the insincere, the resistant, the skeptical, and they have rewarded those who have appropriately suspended all disbelief and espoused the gods' wishes.

So it's not surprising to me that technology, in this case, this teleconferencing technology, had an uncanny nose for the resistant skepticism of the likes of Milton and myself and punished us for our lack of faith, while at midnight or 11:30 P.M. it singled out for reward the whole-hearted belief in and espousal of technology of our distinguished colleague Larry Keeley. This form of divination of technology is of course a subterfuge, and I will return to it. But it is really funny that it would work for those who believe in it. It's almost like healing. They say if you believe in it, you will be healed, but if you resist it, salvation is not at hand.

So this morning you are going to hear from a resistor. I'm going to frame my presentation in terms that will make nostalgic references to the past.

This is the context in which I want to present my talk: *Quo vadis?* Where do we go from here? It's really interesting that our present culture, civilization, and technology, together, seem to have made the present the starting point and the future the destination. The true meaning of *quo vadis* should include a comprehension of how we got here in the first place. If you have no understanding of how the present was shaped, it's going to be a rather shallow enterprise—figuring out where we go from here.

Unfortunately, or fortunately for some of us, the past cannot simply be ignored, destroyed, or dismissed with a simple statement. It's as if the questions of where we're going to go with our technologies, where design is headed, where design should be headed, somehow appear in a historical vacuum. If I achieve nothing else this morning, I want to invite you to reflect on the ways in which some kind of historical perspective—don't be afraid I'm not going to tell you history stories—can be restored to us as individuals, as citizens, and as designers. Because without a historical perspective, what you will get are disem-bodied pieces, glimpses of information, hope, fear, anxiety, and oppor-tunism. You will get pieces of things that don't coalesce.

The necessity of finding a way of restoring the past to its important place is not in itself hostility towards the future. It is not a way of clinging to the past. Curiously, what happens is that when you ignore it, the past rears its head consistently. The paradox is—and let me start just from one place—that those who seem to thrive most on change and chaos, on invention, re-invention, and innovation, are themselves extremely dependent on orderly and stable phenomena. We want, always, to have our cake and eat it too.

Take a new corporation, the new giant, the new IBM of our new global economy, Microsoft. What Microsoft would like, the kind of open, laissez-faire economic and technological context in which it can thrive and excel and show its mettle also requires the kind of stability and orderliness without which the wealth it accumulates is essentially worthless. It needs an orderly society, a police force that can impose the laws, an orderly stock exchange in which the shares held represent value, and the dependability of an infrastructure, including the roads, the water lines, and the flushing toilets, which come from old and stable technologies that make life possible.

So it is rather a crazy idea to talk of chaos as an end in itself when any definition of success, or the transformation of the future, must imply a certain degree of stability, order, and institutional continuity for its beneficiaries. This doesn't mean that every institution needs to be preserved. But society without institutional continuity represents a kind of social and political chaos that cannot be dismissed as simply part of ongoing change. If you think along those lines, suddenly it appears that the question is not whether there should be change but, "What should change?"

The biggest frustration with beginning with the present is the implication that change is totally autonomous, that somehow we are passive spectators, and that as spectators we cannot have any input, to put it in technical terms, or in more fundamental terms, that we cannot make choices, or have preferences. Just as there is the option to act, there is the option not to act. Just as there is the option to execute, there is the option not to execute. The danger of reifying change (which, by the way, has always been going on) is, as in every previous period in history, to assume that one's own experience of change is very fundamental, very drastic.

I know, nothing since the deciphering of DNA may approach the proliferation of what's happening now (the parallel being the way of conveying information, the information revolution); but there have been several other drastic thresholds in the past during which those who were involved assumed that the earth was shaking under them. And when people are caught in ongoing change, there are usually two impulses. One is to say, "It's happening; what can I do? I can get on the wagon. I can catch the moving train, and I'll just go where the train is going." Or some people can say, "Do I want to get on the train? Where is the train going? Is there anything I can do? Is there an alternative?" But in order to think in a way in which we are not passive spectators in the presence of change, we must make the basic assumption that we are capable, and at times obligated, to make choices.

184

The reason I call this morning's talk "The Twilight of Humanism," is that humanism was, and I say *was*, and maybe for some people it still is, a long-term project. It's not something you can fix with dates. Some trace it back all the way to the Greeks, others to more recent social events such as the Enlightenment in the eighteenth century, or the Renaissance.

Humanism recognizes that we humans, both as individuals—hence the individualist tradition of humanism—and as a collectivity, a society, as humankind, have certain characteristics, capabilities, and obligations; and that somehow, humans have traits such as rationality and values such as moral, ethical choices and responsibilities and the capacity not only to desire, but also to imagine. And since there is the capacity to desire and to imagine, there's also the capacity to chose and to shape, through desire and imagination, one's own and the community's future.

Humanism was a shift from thousands of years of mythic systems, in which what happened to individuals and communities, was believed to have been determined elsewhere. People saw themselves as the recipients of the whims of the gods. Humanism tries to say that we have something to do with our fate, not our faith, but our fate. Humanism is the opposite of superstition, the opposite of fatalism, and the opposite of the notion of being the passive receptacle of the whims of the gods. It is a reaction to a religion that basically assumed that individuals had to always obey; that their destinies had been preordained.

Humanism was an attempt to give to each human being the conviction, the belief, the determination, and the moral obligation of being a participant in, and the shaper of, his or her own destiny. That's the humanist project and it mixes together such ideas as thinking, rationality, a sense of identity, the perception of continuity, and, of course, the respect for every human quality that could make a contribution to this project of emancipation, self-actualization, and self-determination.

Now the humanist project, as with any other similar project, has been neither an easy one nor has it been an unmixed blessing. It contains, believe it or not, the seeds of its own destruction, a challenge to itself. The twilight of humanism may very well be deserved. It is a gross arrogance to really take all the power and all the responsibility for our fate. The gods are not dead; they may be either asleep or waiting for us to destroy ourselves. The true humanist must recognize the limits of his or her own powers. But I think that probably the greatest and the most ironic, if not invidious and pernicious, accident of this humanism has been its marriage to technology.

Humanism has encouraged knowledge, and it has specifically encouraged knowledge which has legitimated and institutionalized and encouraged science. And science and knowledge together have been the twins of this humanism; not only the belief that there is a truth, and there is a natural order, but that we are capable, through our reasoning, logic, and activities, of discovering them.

Humanism has attempted to demystify nature, to reduce nature, to dominate nature with the belief that our capacity to know and to control gives us the tools to dominate. The pursuit of knowledge and an anthropocentric vision of the world combine in an almost frenzied way through the sciences. This combination, of course, crystallizes itself in the most concrete way through technology. And the trap, the *hidden* trap of this pact with the devil, is, of course, the trap that Goethe recognized in *Faust*. What happens is that the road companion that you want to take with you in your pursuit of humanism turns out to have a mind of its own.

This pact with what Goethe recognized as the demon, is our pact with technology/knowledge. The pact with technology says that basically we want it. It is the foundation of our power; it is the foundation of our mastery of the universe; it is the tool with which we impose our human superiority on earth, in the universe, the physical world. We think of it as our creation and our companion. But as *Frankenstein* illustrates, you cannot always control your creation.

Now there was a reference made yesterday to the Luddites. Larry dismissed the Luddites, as some people are nowadays called, because, he said, after all, the Luddites had lost. Historically, Luddism is a very specific event that originated in 1811 and lasted for a few years. A bunch of craftsmen, as he said, got together and opposed the introduction of technology in textile manufacture. But he said they lost. I want to tell you the part about how they lost; I just want to make a little correction.

The English government put together, and sent to the three counties in which Luddism occurred, about 16,200 soldiers within an eleven-month period. To put that number in perspective, that is more soldiers than were assembled under the command and leadership of Wellington to fight against Napoleon. The British Empire, the state, the king, and the established powers, used more soldiers to put down the Luddite rebellion than they used to go and fight Napoleon. That means that once you make an alliance with technology and some people taste its benefits, it is not easy to divorce it. What happens is that the state, the established power, becomes the partner and, as some people call it, the servant of that technology and its economic benefits.

Now, I want to make this talk brief, but I want to quote a very interesting line from F. Scott Fitzgerald so that you don't think I'm sitting here, a sour, angry, old-fashioned man who is opposed to what happens through technology. Fitzgerald wrote, "A first-rate intellect," and I assume designers are first-rate intellects, "should be able to see that things are hopeless and yet be determined to make them otherwise."

The reason I have started my conversation with you on a rather pessimistic note, on a resistant note, is to say that for a genuine, open, free, and responsible orientation to the future and an attempt to answer the question, "*Quo vadis?*" in an intelligent way, there is no need to be Pollyannish. There is no need to minimize the risk. There is no need to pretend all is well. There is no need to resign oneself to a passive role. There is no need to accept change in all its aspects and consequences as simply inevitable. There is no need to resign one's moral responsibility. The true expression of what I call a healthy futurism is to be totally sensitive to the traps, the handicaps, and the pitfalls.

Yes, I'm as capable of saying, "Wow!" and enjoying what is "neat" and unique. No matter how much, in a cerebral way, I try to resist the manipulation of my senses, I thoroughly enjoyed the Industrial Light & Magic show and the clips from *Jurassic Park* last night. I was delighted. And I also realized, thank God I had not seen the film, because it should have been seven minutes long. And that's good. That's good. They should have a seven-minute version for adults, for that part of our adulthood which, thank God, does not extinguish the child in us.

Yes, I think we should pursue technology to create wonder, to be curious, to stimulate our senses. Yes, I think we should play with things; play is very healthy. And there is nothing harmful if you realize that somehow you are being manipulated. I don't mind being manipulated as long as there is enough space for me to keep some perspective on the way in which I am being manipulated.

Of course we should make use of technology as a tool, but we should always keep in mind that it is not a neutral tool. That is essentially a self-serving, or a delusional statement—to assume that any tool is neutral. It reminds me too much of the kind of logic involved when people say, "Guns don't kill; people kill." While it's true that people pull the trigger, guns have a function built into them. When the function of an instrument is to kill, its function eliminates its neutrality. Objects, designs, and procedures have functions. And functions are not neutral. Functions are loaded with meaning and intention.

I have avoided giving any talk in Aspen without using what I still consider the most apt way of defining—even though the word *defining* in this case is not correct—the most apt way of *seeing* what design is. "Design," an old colleague of mine said—and I always wish I didn't have the moral compunction to attribute this quotation to someone else—"Design," he said, "is the introduction of intent into events." Design is fundamentally intentional. Without intention, there is no design. You must have intent. You must intend. And intentionality implies some preference, some choice, some motivation to make it shorter, longer, better, more beautiful, cheaper, faster, whatever it is.

If you have no intention, don't take up the pencil or turn on your computer. Without intention design collapses. It is intentionality that gives it direction, that gives it content. It is intentionality that gives it strength to go forward. And if design is intentional, we better own up to our intentions. If design is intentional, neither its tools nor its creations can ever be, or be perceived to be, neutral. There is no neutrality in human intentionality. We are not ourselves neutral animals even though all the talk about evolution would make you believe that somehow we do make up a part of a very undifferentiated whole.

One of the things in the last few days that brought home the point to me with great vigor is how during the twenty-five years that I have been involved either centrally or peripherally with the profession, I have noticed the change in the language of our discourse. It is amazing how over the last twenty-five years certain words have essentially disappeared from our discourse and our vocabulary. I have also noticed how certain words have become more frequently used and from my perspective, misused, and very frequently abused.

Years ago, there was nothing to be ashamed of in referring to such issues as design and justice, design and beauty, design and equity, design and human betterment, design and social responsibility, design and choice. There was a period in which that kind of discourse was not laughed at, was not deemed to be a bothersome externality leftover from some kind of irrelevant, marginal, tendentious, dying ideology.

Whether for us designers as citizens, as professionals, as workers, or as people with tools that can make changes, create environments, communicate ideas, create objects, or invent utility, we wanted to believe that some time sooner or later all that we did bumped into such ideas as justice and fairness, peace and responsibility, and all other issues related to human betterment. That kind of talk was not a measure of soft-headedness or Luddism or reactionary mentality. It's very curious; that language, those words, are either out, taboo, or between quotation marks. There are, instead, other words that have all become much more common, more frequently used, and shamelessly abused.

One of the ones I like the most is how the word *community* has reincarnated itself in cyberspace. Another attribution I cannot ignore, is to a New York poet whose name I don't even know, but who I met at some meeting, and who made a very interesting comment. He said, "The difference between a place and a space is that places are spaces with memory."

Which brings me to the second misused term, the word *memory*. The word memory as in *memory bank* or an electronic system has nothing to do with what we used to call, and what we as humans experience as, memory. An archival data bank that regurgitates information has no connection with what we as human beings experience as memory. Memory is related to experience which is continuously dynamic, never objective, never out of context, and continuously being mediated by changing perceptions of the present and our needs. Memory is fundamentally affective as well as rational. It is subjective as well as objective. Memory is what blurs the line, and yet we cling to it.

Memory is the foundation of our identity, an identity that is fluid, adaptive, and fragile, while the memory we've been talking about nowadays is a memory essentially devoid of any of those characteristics, and utterly incapable of substituting for the subjectivity of experience. We still feel compelled to refer to things as *virtual*. I think the time will come when we are going to drop the word virtual. And when we drop the word virtual, we are going to go back to a notion of reality in which the virtualists will no longer have any memory of the memory I'm talking about.

Now another couple of points, specific references to the kind of conversations that we've had here, which again show the partiality of explanations that are fundamentally self-serving. And I don't mean self-serving in a personal way, but self-serving to the emerging institutions which technology dominates. For instance, I was taken, excited, and compelled by the first night's presentation by Tom Peters. I loved, particularly, that wonderful word *mindfulness*. Nice words: *design mindfulness*. But one has to resist equations in which there are missing variables.

Tom Peters rightfully, and we can all sympathize with him, was against the monotony of boring sameness. He said, with some faith in it, that corporations, like himself, had had it with the commoditization of production; that somehow making a commodity of production creates not only boring environments and boring products, but ultimately starts eating into the profit margin. I cannot disagree with that. Let's for a moment try to find out why.

Why does this commoditization occur? And then we realize where the hidden taboos, the hidden secrets, are in the closet. The commoditization of production happens because labor is becoming a commodity. You cannot get a product that is not a commodity if the laborer that participates—and I mean by laborer, the worker, including the designer, the engineer, the inventor—if the person making it has himself or herself become a commodity. Just like the rest of us, business wants it both ways. They would like the worker, the unit, the laborer, the inventor, the creator, the designer, the artist, to be transformed into a commodity, but they don't want the product of his or her labor to be a commodity.

When the craftsmen were making it, the work was not a commodity. That's because craftsmen are not commodities. In our complex business system, the first and foremost commodity is the workforce. It's a commodity because it is reduced simply and only to its production value. The transformation of the person into a utility in the production system is the very definition of commoditization, because the moment that same utility can be achieved by another person it is substitutable. It is the erasure of the distinctiveness, the uniqueness, the singularity, which we decry in objects. It is not simply that Toyotas look and act alike and are indistinguishable from Hondas. The engineers and the bureaucrats of Toyota are also indistinguishable from the engineers and the bureaucrats of Honda.

The commoditization happens at the level of work and not simply at the level of the object. The work precedes the object. We must stop making secondary, clever, insightful observations and refuse to dig deeper. First and foremost the designer must dig deeper. You have no greater challenge than to challenge your own assumptions, particularly the assumptions that underlie the statements of every bloody guru, including myself. If you don't challenge that assumption, there is no design.

Yesterday I was telling someone that we think designers are people in search of solutions. But do you think it is necessary that designers be in search of solutions? That designers put forward first and foremost a solution? How can you say the solution is primary, when the first thing that designers must always do is decide whether the problem is worth solving? How can you assume that somehow that question does not remain relevant, all along, from inception to conceptualization to execution to implementation; that that question does not at all stages accompany you? The worthwhile-ness, the relevance, the significance, the impact, the consequence, the utility, the foreseen, the unforeseen, the short-term, the long-term, the medium-term: these are nothing but questions. The future is essentially a question. The future is an invitation to questioning. The future is not an alternative garment to put on simply because it is handed to you. The future is not handed to you.

Quo vadis? Well, guess what? I will simply ask you the question: Where do you want to go? Not where you are going, or where will you go, or where will you end up, or even better, where are they sending you. Please let me ask you: Where do you want to go? *Quo vadis?* The humanist project does not present the question of *quo vadis?* as the activity of a person looking into a crystal ball to figure out what the future portends. Humanism tells you not to look into the crystal ball, but to essentially engage in the most profound exercise of all—the self-examination that challenges your own choices, your own desires, your own weaknesses, fragility, frailties, preferences, and your own so-called moral center.

In this era of coining new words, we are making great slogans, sentences, and buzzwords. There's a new book called *Cyberia*. Has anybody read that book? With *cyber* you can start any word you want, cybernaut, cyberspace, cyber this, cyber that. Do you know where the word comes from? People ignore this; it is a Greek word meaning *helmsman* or *governor*. It's very interesting. The Greek word that somehow got into cybernetics is one that refers to guidance and governing. It's curious. Somehow you realize who is the helmsman and who is the passenger. It may very well be that the Greek term for helmsman is an ironically apt metaphor for cybernetics. They didn't have governors; they had helmsmen. Helmsman of the ship.

I want to coin my own word to throw into this morass of word-mongering and wordsmithing. The word I want to coin is *technonarcissism*. I think it's a good one and I'm not ashamed of it. I was trying to be clever about it, and I first said, "Well, you know our kind of technology is the bastard son." My very first lecture as a university graduate student was to four hundred undergraduates. I was giving a lecture on the French Republic, the Fifth Republic, and I wanted to get them going, so I said, "The French Republic is the bastard child of DeGaulle and Brigitte Bardot." Oh, my God, the undergraduates loved it. It's funny how the first opening sentence of your very first lecture gets you applause and you're stuck in that metaphor forever. That's what's called reinforcement, learning by reinforcement.

So here, thirty-two years later, as I was trying to come up with images, I went back to this imagery of bastardy. I said that the most recent version of this technology is the bastard child of a necrophilic father, war and defense and militarism, and a rather promiscuous, whore-like mother, entertainment. They have, for the last twenty-five years, been the two most powerful forces driving this emerging technology that is transforming, in one way or another, all of our lives. This technology is essentially the product of an unholy marriage between the instruments of death and the instruments of passive entertainment—with all due respect to the much-abused term *interactivity*.

Technonarcissism in the nineties is similar to the child of a narcissism of the seventies, the "me" generation. Remember that? We had a "me" generation. Well, it's not totally dead. After the yuppies and the X generation, the obsession with the self is still very well and alive, a self-isolated, self-absorbed, self-indulgent self, cruising lonely on the superhighways without touch, contact, feeling, or hurts. Not that I'm highly touchy-feely, but the "me" generation had that aspect. We dropped that part, but kept the narcissism: The adolescent incapacity to genuinely connect is a leftover from the seventies. You know what we had in the eighties. It was the generation that said, "Greed is good." When you marry the seventies to the eighties you get laissez-faire, anti-communitarian, anti-public responsibility, unaccountable, blind faith in change for its own sake. But as Fitzgerald said, If you're a great intellect, in spite of the bad news, you still have to change the world.

Designing for Democracy in Cyberspace
Mitchell Kapor

By way of introduction, I just want to say I found myself in surprising agreement with virtually everything Jivan said, since Jivan doesn't appear to like technology, while I'm going to approach that same critique from the point of view of someone who basically likes technology and feels comfortable with it and attempts to embrace it. Let there be no mistake, I think I am as much a believer in the humanist project as Jivan is, and I see the extreme challenges that it faces now.

There are some interesting, difficult challenges about how to proceed ahead. Not the least of which is the extraordinary way large corporations have co-opted ideas in this Postmodernist era. I was reminded of this by the question, "Where do you want to go?" which Jivan so eloquently raised. Did anybody else notice that this is the major tag line of Microsoft's TV and print campaign? "Where do you want to go today?" It's amazing to me how the most profound questions can be co-opted in the commercial sphere.

I also want to say that there are different varieties of technophiles, and I need to distinguish myself from some of my colleagues at the MIT Media Lab and elsewhere. There is a notion currently running around which you may find hard to believe—but trust me, it is taken very seriously in certain high-tech quarters—which is the idea that it would be a desirable thing to be able to upload human consciousness into a machine to preserve it and thereby cheat death. As the first generation of pioneers in artificial intelligence is reaching retirement age, this project has taken on an added urgency. What I find striking is that even to think that this is something that is possible requires a kind of flattening out of memory to information retrieval, of the type Jivan was talking about, that is to me quite unthinkable. And the whole notion of being able to cheat death is one that is fundamentally at odds, not only with the humanist project, but with my own Buddhist point of view that some aspect of suffering is fundamental to our condition. There are technophiles and there are technophiles. And I'm one, but I'm not one of those.

194

Well, I want to tell you a story, to frame what I'm going to talk about. This is something that took place in my class at MIT. I just finished teaching a seminar, this term, on democracy and the Internet, which has had a lot of topicality lately as we're seeing the imminent passage of a bill which will fence free expression on the Internet. My concerns were to actually go back about five years to the founding of the Electronic Frontier Foundation.

We were having this seminar, and what we were discussing was social migration in cyberspace. If you're willing to go along with the spatial metaphor here, I'd appreciate it. Either you believe it or perhaps I could ask you to suspend disbelief. But fundamental to the metaphor is the notion that cyberspace is a nonphysical kind of space. It's the place two people are when they're talking on the telephone. As my colleague, John Perry Barlow, says, "It is also the place where all of your money is except that which you might have on your person at this very moment."

Where are these social migrations? Well, first, there's been the Internet, which has grown up over the past twenty-five years. It started out as ARPANET, a little Defense Department project that brought together researchers in computer science and other academics and a few intrepid digital pioneers. It has become, over time, its own realm with its own customers and traditions, and ways of speaking. And the independents have come along in the past few years, consumer on-line services—Prodigy, America Online, and Compuserve—which for the first several years of their existence were island continents unconnected to the Internet. You could not get from one to the other.

However, with the growing popularity of the Internet, America Online management began building bridges which enabled their users not only to exchange electronic mail with people on the Internet but also to participate in discussion groups, computer bulletin boards, computer conferencing—services that were on the Internet. And so there was a huge migration, when the bridge was built, because there's no political control or customs or passports. Hundreds of thousands of people took advantage and did show up all of a sudden in these various discussion groups.

This is what we were talking about in class, when one of my students, a very bright student, said, "You know, I was on-line the other day, and I was in a news group, a discussion group, and a new user came on and I could tell he was from America Online. You know in cyberspace you don't have a body but the way you can tell is that you do have an electronic mail address. And your electronic mail address contains where you come from: "aol.com" indicates an America Online user. And I knew that before he even opened his mouth, so to speak, he was going to say something absolutely clueless because that's the way AOL users are."

Now those of you who are on-line know that that is a very typical sentiment. I said to him, "Now wait a minute. Let me see if I've got this right. You don't know anything about this person except one fact, which is that he's on America Online. You have a belief that all AOL users are clueless. This person is an AOL user; therefore, he is clueless. Tell me something. If you were doing this in the real world, what would it be called?" Everybody got it. I mean, it would be something like racism or sexism. It would be some kind of "ism," and my own contribution to coining new terms is to call this *domainism* because your domain, your electronic mail address, is the visible characteristic that you cannot hide in cyberspace. It turns out that people will use whatever information there is, however negligible its real content is, as a basis for making judgments and acting.

I think this incident illustrates what I call the first noble truth of cyberspace, which is that we bring our baggage with us. All of the ways that we behave as individuals and as society, all of those things are going to show up in cyberspace. Cheats and crooks and criminals will find their way onto the Net and begin to carry out their scams. There's no escape. The reason I think this is a salient point to make is that one of the great myths about cyberspace is that it represents, or is, a kind of electronic frontier where a fresh start is possible—where one can leave behind the limitations, the boundaries, the sinful cities and return to some primordial condition of freedom and openness and possibility.

Now, since I'm one of the principal people that helped construct and promote that myth of cyberspace, I feel that it's only fair that I should take my turn in deconstructing that myth. That story illustrates very well, I think, the point that there really aren't fresh starts. We're going to bring ourselves to whatever party we go to. But there are some other aspects of the myth of the frontier that have been applied both to the physical frontier of the historical West and are widely applied to cyberspace, that need some comment.

The one that I want to draw your attention to is the myth of rugged individualism, this notion that somehow the West is a place where "a man can be a man." Let me not even comment on the implicit sexism in that, but call your attention to the fact that somehow the frontier is thought to be a place in which one's fundamental personhood is more readily capable of expression; that it is simply the presence of some kind of external constraints which precludes us from the exercise of our full freedom.

If you look at the history of the American West, you find out that the most profound examples of rugged individualism were the undertakings of the robber barons who built the railroads, the capitalist swashbucklers who let no object stand in their way in order to create and develop economic empires. They did not restrain themselves when it came to any form of exploitation, up to and including outright violence, in the pursuit of their individualistic aims and goals.

In the century or more since the closing of the frontier, I think, the cultural baseline of acceptable behavior even for robber barons has risen. A rising tide lifts all boats. But it is for that reason, I think, that we don't see the Bill Gateses of the world hiring thugs to go kill protesters the way the robber barons of a century ago certainly did. But we see a similar style of exploitation of resources, a certain will to dominate and to control, that characterizes and typifies the motivations of the heads of large organizations as they encounter this huge new frontier.

Now don't get me wrong on this. This is not an overall indictment of business or of capitalism or of the free market. Over the past several days there have been a lot of well-thought-through and passionate evocations of the need for creativity and dynamism and innovation within business, but I think that business also has to be held account-

197

able for its moral behavior or for its lack of moral behavior. To ignore those questions and issues is to paint a very one-sided and superficial view of business.

If we're willing then to leave behind the myth about the frontier of cyberspace, then what? Where does that leave us? Well, I think cyberspace is as urgently in need of a new ethos as everywhere else in society. That's a pretty tall order and one that I'm not sure that I'm up to. But I want to just leave a kind of link to a hypertext that I can't fully talk about today as to what might be involved in that new ethos. I want to move over that pretty quickly, because what I'd like to spend the rest of the time talking about is the role of design and designers.

It seems to me and it was very well put just a few minutes ago that somehow we need to reclaim for the digital era and for a Postmodern era, the traditional goals and values of humanism, of justice, and of equity. We need to do that if for no other reason than that we simply can't afford to have this era of Postmodernism owned and controlled by people who are fundamentally nihilistic. The problem that those of us who have that agenda face is somehow finding a way to claim or reclaim a share of the moral high ground from conservatives and from the radical right in particular. It has become unfashionable if not impossible for people of a liberal bent or persuasion to talk about the importance of issues of character, personal responsibility, and culture without being accused of selling out to the right. And that I think leaves liberals weaponless.

So I will cap this point off by suggesting that you look for a book called *Emotional Intelligence* by Dan Goleman, who is a science writer for the *New York Times*, that is an attempt to reposition character traits such as perseverance, tolerance, self-restraint, and altruism not as the captives of any ideological point of view, but as parts of what it fundamentally means to be human, and as part of a kind of developmental agenda.

The title of the book, *Emotional Intelligence*, refers to the idea that just as there can be a curriculum for children to develop their cognitive intelligence, meaning logic and mathematics, we could begin to think in terms of a curriculum for the development of emotional intelligence.

And he reports on some very interesting psychological research that backs this up. So think of what I've just done; if you were reading this on a page in the World Wide Web there would now be a phrase in blue called *emotional intelligence*, and if you were driving rather than listening to me, you could click on that at this point and go off into that subject. In fact that's an advantage of hypertext as a medium. It empowers the user in a way that this type of medium doesn't, because unfortunately I have to make the choices.

What I want to talk about is the role of design and designers in cyberspace. Here I owe a great debt of gratitude to Bill Mitchell who is the Dean of the School of Architecture and Planning at MIT and my mentor there. He has written a wonderful book which has just come out from MIT Press called, *City of Bits*, from which these next few points all derive originally.

If you think of cyberspace as a kind of nonphysical place where people hang out and do things—they might shop, they might chat with other people, they might engage in work with others—then the traditional set of questions that urban designers who are actually involved with the design of real cities are concerned with are highly germane. How do we make this a good place to live and to work? What arrangement of the parts—what facilities ought there to be in order to help people be more human and go about doing what they're doing?

Urban design, which I know next to nothing about other than having taught a course on digital communities with Bill Mitchell, looks at that in the context of cities and asks questions about access. If there are public places, is there access to them? Genuine public access that is inclusive? That everybody can get to? Is there a transportation system that is inclusive? Or is a place somewhere where you have to have a car because there's no public transit? So there are an analogous set of questions to be asked about cyberspace. Are there public places or not? Does cyberspace consist of an endless series of digital shopping malls and digital Disneylands?

Nothing wrong with those taken one at a time, but it's a fair question to ask what is the entirety of the environment? And what is the nature of the public places where people can come together? I will come back in a few minutes to look a little bit more deeply at those issues. But the point that I want to make is that design questions like these are fundamental to cyberspace, and every time someone is engaged in the project to build one more page on the World Wide Web these questions and issues are lurking there. The questions that one might well want to ask as a designer are the questions that Jivan raised at the end of his talk.

One thing about design, though, is that it always involves a particular medium, or media; so graphic designers work, among other things, on paper with print and with type; architects work in the medium of buildings and three-dimensional human-sized objects; software designers work with the medium of construction of software or computer programs. And so designing for cyberspace raises the important question: What is the nature of cyberspace as a medium? What are the characteristic constraints and affordances that are present when people sit down at computers that are connected over networks and interact with one another?

That is a whole topic unto itself, but I want to focus on what I think is one extraordinarily important aspect of cyberspace or the Net, which is that of decentralization. Compared with other information and communications media, cyberspace and, really, to be more precise the particular information and communication media that are built on top of it like electronic mail and computer conferencing and MUDs and MOOs and other exotic sorts of things all operate in a very decentralized fashion.

Someone once said that the Internet is the world's largest functioning anarchy. There is no Internet Inc.; there's is no one central party that is in control, that sets the standards, that decides what new features and capabilities will be added. It's not like the old Bell system. In fact, it is the exact opposite. It's a system which manages to operate today because each of the tens of thousands of different networks which comprise the Internet has agreed to operate on the basis of the same technical protocols. So that the fundamental operation is through cooperation not coercion.

That is historically unprecedented for a major medium. Certainly television does not work that way. With television you have a relatively small number of providers of videoconferencing, if you include significant cable networks. (A relatively small number used to mean a half a dozen now it probably means about fifty.) They exist in a highly asymmetrical relationship to their viewers. They have central control over what you see, when you see it, and what they think is important. They define that reality. They have a legal monopoly because of the broadcast licenses and because of the local monopoly on cable. There hasn't been anything much that anybody could do about it. It's a centralized medium.

On the Internet, on the other hand, anybody with the requisite technical knowledge and a few thousand dollars—so that doesn't mean 100 percent of the people, but it means millions of people and institutions—can create their own home on the Internet; they can put out a server; they can begin to publish information on the World Wide Web; they can have a voice. There isn't anybody to say, "No you can't do it," or "You must do it this way."

That is encouraging enormous diversity and enormous experimentation and an enormous amount of noise and garbage. The rule says that 90 percent of everything is junk. But I'll be the first to admit that in singing the praises of diversity, it comes at the price of being willing to wade through or having a way of filtering out the mundane, the boring, the unimaginative, and the repetitive. But if you go out on a day-by-day basis on the Internet and in particular on the World Wide Web, there is a fantastic explosion of creativity, and barriers that many people thought and many people still think could not be broken are being broken.

There's a kind of information lag as to how rapidly the rest of the world finds out about this. I'll just give you one example. There is a new technology called Real Audio™ that is now available on and through the Internet that makes it possible to deliver audio, mostly speech but also music, to anyone who is connected to the Internet, not over a high-speed line, but over an ordinary phone line; and not in a way where you have to wait fifteen minutes to download a two minute audio clip, but in a way that two seconds after you click on whatever it is, the soundtrack begins and delivers you a real-time audio stream. So it is not only what radio does and can do, but it goes several steps further because it enables the users to get at the information when they want it.

So if you listen to NPR—and I suspect there are a few NPR listeners here—if you don't happen to be around when Terry Gross does "Fresh Air" or whenever there's a show on at two o'clock in the afternoon, you're out of luck. But NPR has a real audio service, and they're beginning to experiment with it, and you can get to the interview that you wanted when you want. But not only that, if you wanted to set up your own radio station on the Internet, as it were, you could do that. No FCC license required. This is what people are doing and they're doing amazing and creative things.

So the Internet is not just about text, it is also about other types of media. Again you will hear it being dismissed in the following terms, "Oh, that's toy stuff. You can't get the kind of sound quality that we can do in professional audio." And, in fact, that is true. The quality is as barely adequate as was the processing power of the first generation of personal computers, which is to say good enough for the first few hundred thousand or million people to buy spreadsheets and word processors, and on a trajectory that will absolutely and certainly overcome the performance and the quality limitation. Not here yet but coming tomorrow.

So the point here is that when you have a new medium, or actually a meta-medium, that is fundamentally decentralized in its operation, that has very low barriers to entry, that encourages innovation and experimentation, you can expect an enormous explosion of new products, new services, and people trying things out. So some very important questions then arise. What kind of places in cyberspace are likely to emerge?

Well, there is already a lot of cyberporn. There are more cyber shopping malls than the economy of cyberspace probably needs. The first few cyber casinos are gearing up. They're off shore; they're American companies that have moved to Belize. They're going to let you gamble from the privacy of your own home and bet on the horses and play blackjack and craps and roulette. And some of you here, I venture to say, are either working for the cyber malls and the cyber casinos or will have the opportunity to do so, because they're going to need their spaces designed and they're going to need graphic design and interactive design and user-interface design and so on.

And the question is, as I said before, is that all they're going to do? Don't know. And in my more pessimistic moments I tend to think that the bright dreams that people have had for cyberspace as a place that would encourage diversity, encourage new kinds of expression, encourage or give more of a voice to ideas at the margins for which there is no room in the mass media, that those hopes may be crowded out by something which is very flashy and provides extraordinarily intense but ultimately superficial experiences.

Here I'm thinking of the day, which is not too far off, when the technologies of virtual reality and the technology of the Internet become narrowed. Will that become the mainstay, stock and trade, of the Internet? Or will we find ways? Will we find commitment and the capital to design the equivalent of town squares and cafés, third places that are neither home or work that encourage association and the coming together of people for common purposes? And that last point is a very important one because there has been an obvious decline in the ways in which Americans get together.

There's a wonderful article in the *Journal of Democracy* called, "Bowling Alone," by Robert Putnum, who's at the Kennedy School of Government where he looks at the different measures and surveys of how much people are getting together in a voluntary and associational way. The statistics are pretty gloomy. One of the statistics from which the title comes is the fact that bowling itself is more popular than ever, but participation in bowling leagues is down. In a given year in this country, some 80 million different people go bowling. I don't know what the weekly figure is, but it's a very sizable portion of the population. Yet participation in bowling leagues is down dramatically. Which is to say, people are going bowling alone, and they're doing lots of other things alone that formerly used to take place in groups.

I recommend your attention to the article, because the reasons for it are pretty complex and not always agreed upon. The reason I cite it here, though, is to ask: Can there be meaningful ways for people to get together in cyberspace? And I'm not talking about replacing face-to-face meetings with some type of virtual encounter. I'm talking more about supplementing or augmenting what people do with each other to reflect the fact that we live in ways that are perhaps physically isolated and separate from the people we'd like to hang out with.

In the audience, I think, is Stewart Brand, who is the founder of the WELL, a computer conferencing in the San Francisco Bay area, that has convinced me beyond a shadow of a doubt that *real virtual places* can be creative and maintain a sense of community. What do I mean by real virtual places? Virtual in the sense that the WELL as an experience, while tied to the Bay area, is really in cyberspace. But real in the sense that the kinds of connections, relationships between people, experiences, and what people get out of it are as important to them as what they do in the rest of their lives, at least when it is at its best. But it is a kind of existence pool in which community is not limited to the physical thing.

Having said that, I will caveat that extremely heavily in the same way that Jivan did and say that *community* is an incredibly abused and overused term, the real meaning of which we have largely lost sight of. If the WELL exemplifies cyberspace at its best, it also raises enormous design questions that have barely been explored. How do we make a good virtual place? What kind of environment, technical facility, user-interface, and style of interaction contribute to that? And what kinds of elements make it more difficult? Once you've built that kind of environment, what style of interaction, in other words, what cultural norms or cultural practices of the inhabitants tend to foster preservation and which foster its decline?

Anybody who has spent any amount of time on-line knows that the minute that there's a discussion in cyberspace a flame war is not far behind where the conversation sinks to the least common denominator. Somebody accuses somebody else of being a Nazi and everybody gets turned off. We don't know too much, to be honest, about how to help manage and facilitate conversations so that they do not degenerate into flaming. What I'm really suggesting, then, is that there are an unbounded number of challenges for designers to undertake in cyberspace. First to understand, and then to do. How do we build environments that are more deeply reflective of our human needs and wants and loves? And so we have to start by establishing that as a legitimate concern and a legitimate domain of discourse.

There has been at Stanford for a couple of years an interesting course in training software designers built upon the studio method, the same way that graphic designers and architects have been trained for a long time. But it is totally unprecedented when it comes to computer science, the idea that you sit people down in a studio, people who are there to learn with people who are experienced, and that you do hands-on work, you tackle projects, you learn by doing, you get feedback, you get critiques.

We're taking that studio methodology at MIT this fall in the School of Architecture and Planning and offering a software design studio out of the conviction that if we're going to meet some of these challenges, we need to train people in design in this particular domain and subject matter, because you cannot be a cyberspace designer if you do not come to terms with the technology. It doesn't mean you have to be a technologist or a programmer. But you cannot look at it and say, "Oh my god, there's all this computer stuff. I just can't deal with it," because that fundamentally disempowers you from being an active participant in the process.

It's a bit difficult for someone who's a mid-career professional who may have come of age, as some of you undoubtedly have, before computers were commonplace. That's a kind of a retraining problem, and there's a lot of fear that has to be overcome. On that point, all I can tell you is just yesterday I spent forty-five minutes unsuccessfully trying to reconfigure my son's IBM PC, fiddling with its config.sys, autoexec.bat files in order to run a new game that he had bought. I couldn't understand why it wouldn't work. I followed the instructions. I tried calling the hot-line, and I still could not make it work. If I have a hard time with it, which I genuinely do, fill in the blanks.

With respect to a new generation of designers who are coming of age and who are just learning the discipline, the perspective, and the practice, I think there is every opportunity to train people who can meet the technology on its own terms and who are fluid with and fluent in it so that when they actually go and do the design, they hold their own with the programmers. I think that is clear.

The responsibility to make cyberspace a good place to live and work carries with it the implication that as a medium we need to understand its constraints and its affordances. We need to be clear about our agenda, because when we design an artifact we're doing so to make something which is suitable for human needs and purposes. And that's why design isn't engineering. Engineers by and large take this issue of what the needs and human purposes are and try to get that out of the way at the beginning—that's the messy, squishy stuff—so they can get to the engineering part.

Designers want to grapple equally with what the human needs and purposes are and how to wrestle that out of whatever medium they're working in. But we'd better be clear about what human needs and purposes and goals we're going to be serving and fostering, ones of consumerism or ones that speak to serving people as citizens, not merely as consumers. So we need to think about that very clearly. And we need to think about incrementally reinventing the design professions to be full and equal participants in society's task of crafting and shaping this new cyberspace. This will, in fact, keep all of us busy for a long time, so I won't take any more of your time today because we've all got a lot to do.

Designing the Infinite Game
Charles Hampden-Turner

Let me first turn to some of the things the earlier speakers touched upon this morning. I think Jivan gave us a good example of humanism with hubris. Of course, the early humanists knew better than to boast; hubris or pride was what caused the hero to fall. And Hephaestus, who was the god of technology, the god of method, was the only imperfect god on Mount Olympus. Hephaestus was the god who limped. If you carry a gun, you can't make love. You can't show nurture and caring or concern for people with a gun; there's only one thing it's good for and that's threatening or killing. And the Greeks knew that and they knew that well.

I'd also say that rugged individualism was a thing that the original humanists did not believe in. In fact, nearly all the rugged individualists came to grief. The original idea of the Greek soul was *psyche*, and soul meant *connection* to the world. Soul was when you poured a libation to the gods, and there was an arc of liquid between you and the god. The connection between you and beauty was the libation poured to Aphrodite. The connection between you and truth was the libation poured to Apollo. And the Greek gods in that form were not supernatural gods as is the Christian God. They were forces in *this* world: the force of truth in this world, the force of beauty in this world, the force of technology in this world. And people had soul in as far as they sought and achieved connection.

Today I wish to speak of this connection and disconnection, as *infinite games* and *finite games* respectively. My view is that we are beset by finite games, win-lose contests that so threaten our pride that we have no choice but to join them. By "we" I mean the English-speaking world of competitive individualists, the heirs of Adam Smith, who wrote *An Inquiry into the Wealth of Nations* in the year in which the Declaration of Independence was signed. I do not include in my strictures the burgeoning economies of Southeast Asia. Indeed, I shall be arguing that they know something we once knew but have lost sight of.

We watched a finite game develop just this morning as Jivan teased Milton and Milton teased Jivan about putting on weight or something. And almost immediately they found themselves pulled into a finite game. They started competing.

The antithesis to the finite game is the infinite game. The distinction is, alas, not mine but that of James P. Carse. In his book *Finite and Infinite Games* he started something which others must continue. This is my contribution to the continuance of the infinite game. The concept is somewhat elusive. I hope it will emerge from the mists during the course of this presentation until you see it clearly.

Let me first explore the game metaphor. Why do we "play" at all? What is the secret of what Johan Huizinga, the Dutch historian, called *Homo Ludens: A Study of the Play Element in Culture*? Certainly in business we cannot escape from game metaphors. How many times have we heard calls for "a level playing field," or listened to warnings about the government "picking winners?" We all want "an even break" and complain when authorities "move the goal posts." To listen to Micky Kantor you would think that capitalism had been invented and patented by Parker Brothers and that the rules had been written by a Puritan saint inspired by a divine CEO.

But why play games at all? Why not be deadly serious? Why games and not battles? I believe that we play whenever we want to *learn vital lessons without being seriously hurt*. A game is a simulation of something so serious that it is wise to play first. Designing is a form of play.

I consult to an insurance company worried sick about ever-rising premiums. Why not create a driving simulator? On that driving simulator, why not simulate all the hazards a driver would meet in a thirty-, forty-, or fifty-year career driving a car? Simulate a breakdown on a foggy road. Have a child run out after a ball. A whole lifetime's experience could be distilled within three hours, and advice could come up saying, "Mrs. Briggs was killed on this corner in 1988. She too met a truck on the blind side." And they'd say, "Take four drinks. Go on take four drinks. Now see how your reaction time slows up. The old lad took four drinks. He's dead. He was killed in this place by a truck of this design just at this very corner just three years ago." If people could do that sort of thing, if they could be put through a driving simulator, then there would be fewer accidents on the road. So I see simulation and game-playing as a way of saving people from some of the rigors and the terrors and the dangers of real life.

Perhaps the earliest form of play in our records of Western civilization was Greek tragedy. This simulated lethal human conflicts in which heroic characters, sympathetically portrayed, destroyed the values they most cared about, including each other. Medea does not really kill her own children. She pretends to, on stage, because she is so furious with Jason, their father. And the audience surely got the message that quarreling parents kill the lives of their children by degrees.

All Greek tragedies have the same structure. The finite game of feuding heroes and heroines proves itself so deadly that it kills the infinite game of parental nurture, royal succession, wise governance, and the like. The horrified audience feels revulsion for the polarized values, pushed to excess, for the process of taking ideas to their logical extremity, a slow dance on the killing ground.

But consider the effect upon the audience of that time. My late mentor, Rollo May, told me that audiences in the Greek amphitheater sat shoulder to shoulder, touching. As the plays reached their climax, a great cathartic shudder, an Orphic vibration, ran from person to person in great waves of empathy and grief. For surely the message was plain. The finite game is over. Everyone is dead or heartbroken. But you, the audience, are alive, and the vibrations you feel are part of the infinite game, the harmony of Dionysus. Put your faith and energy not in the polarities and stereotypes of lethal conflicts, but in values finely fitted together in *symphoniasis,* the ideal of harmony.

It is over two millennia since the golden age of Greece, and still its character eludes us. If anything we are headed in the wrong direction. In Hollywood, Douglas and Stone make love until she reaches for her ice pick—the infinite game interrupted by bloody finality. In genuine, classic tragedy, the phoenix of the infinite game rises from the ashes of the finite game, like the rainbow after the flood or Easter following the crucifixion. As the finite game crashes towards its doom, we reconceptualize, we feel reborn, radiant, like Ebenezer Scrooge waking from his nightmares on Christmas morning. It is not too late, after all. The game goes on.

Human beings are not the only creatures who play in order to learn and in order not to get hurt while learning. Animals play. Tiger cubs, bear cubs, dolphins, otters, and much of the animal kingdom learns to fight, hunt, run, and compete by playing. It looks on the surface like a series of finite games, stags butting antlers in ritual conflict, but look deeper and the infinite game becomes apparent.

Consider the head-butting of stags fighting over their mates. It works out so that the stronger, more agile males pass on their genes. That is the infinite game behind the finite games. But it is crucial that these contests are play, *not* lethal conflict or even wounding. Consider ten stags, with levels of strength ranging from ten to one. If they hurt or killed each other, ten might kill nine, so that nine did not reproduce. Five might wound eight, so that seven would finish him off. In this case, the weakest stags might reproduce. The logic of the infinite game which keeps the species strong is wrecked. *Only* if play remains non-traumatic can evolutionary selection work.

A few years ago, Robert Audry popularized *The Territorial Imperative.* The meta-message of the book was that animals fight for territory, so it is natural for us to do so also. Needless to say, he missed the point. Animals of the same species stake out territory so that they can spread themselves out over a sufficiently wide environment so that none need starve. Too many predators, in too little land, weakens *all* of them, threatening their whole species. Mock-fighting for land is another key to co-evolution, another way of playing while surviving.

But the bulk of my remarks today are addressed to the economy and management culture of the U.S.A. My theme is that our competitiveness is too lethal, too traumatic, and too cruel. We are in grave danger of sacrificing the infinite game, the larger social reality which joins our rivalries together into an overall process of learning and improvement. Our fierce fighting is wrecking the infinite game. Our corporations downsize, threatening employees with the extinction of their life-chances. Millions, younger than I, are already on the scrap heap, unemployable and waiting to die. What remains of their lives is often without direction or purpose.

I have travelled extensively in Southeast Asia, where economies are booming and vitality is palpable. I have consulted extensively to Motorola, one of America's most admired companies and an example to this nation of what is possible. Motorola's annual growth rate has been over 20 percent. It grew 52 percent in Europe in 1994. Motorola, Hewlett Packard, and in some countries Intel, have sensed the infinite game, that heartbeat echoing beneath the surface manifestation of corporate head-butting. Superficially, you see, there is little difference. Their jousting looks like our jousting. Their vocabulary is largely borrowed from ours. They read the *Harvard Business Review* and borrow its slogans. Yet, *the context is different.* It is a difference, I submit, which makes *all* the difference. Listen and you will hear the sound of different drummers. Let me make some comparisons.

The Finite Game Cannot Change
The Infinite Game Must Change

Because of the dominance of the legal tradition in the U.S.A., because it is the only remaining superpower, and because America has assimilated millions of immigrants from all over the world, it has a tendency to behave as a world referee. Listen to Micky Kantor and you get the impression that there is only one legitimate form. We know what it is and everyone else is trying to cheat.

The truth of the matter is otherwise. Capitalism is a game which we redesign as we play. The rules *must* change. Indeed we are stuck in a zero-sum game of winners and losers *unless* the rules keep changing. We accuse the Japanese of cheating when they do not make at least a 6 percent profit. But who says that every company must make short-term profits rather than going for market share with profits shaved wafer-thin and then profiting long-term? We are using our power to try to freeze the rules. It won't work any better than Prohibition, which was a law aimed at America's more recent ethnic immigrants—beer-drinking Germans, wine-drinking Italians and Latinos, and the Irish sodden with native brews.

Because we invented economics—the British and the Americans—we have deluded ourselves that its laws are immutable and stand for all nations everywhere, that all nations will play on *our* level playing field. Well, I have bad news for these "globalists." You will have difficulty buying Adam Smith's work anywhere in Asia, outside the Indian subcontinent. You *can* get the works of Frederich List, the nineteenth-century architect of German unity and of catch-up capitalism.

Some years ago, Bruce Scott revealed the infinite game that Japan, Korea, Singapore, Taiwan, and others have been playing for the last twenty-five years. On the surface they appear to be the same as us, fierce competitive rivalry for markets between individual companies. Yet beneath the surface something else is going on, a Learning-of-Knowledge Race is being prepared. All these cultures have been deliberately running down their products with *low*-knowledge intensity and deliberately increasing their products and services with *high*-knowledge intensity. When I speak to Chinese managers in China, Malaysia, Singapore, and Taiwan, they are all looking for "high-end" work, as they call it; work that will educate everyone who designs, makes, distributes, and uses these products, changing them in the direction of higher complexity.

Notice that this leaves the laws of economics unchanged on the surface. Companies still compete on the open market, still win sometimes and lose sometimes. The change is subtle but devastating in its long-term impact on the West. They are gradually taking over the complex, highly skilled jobs, leaving stagnant pools of low-wage, low-skill work in Western nations, "forcing" us, as we see it, to fire employees and then pay them to do nothing on welfare or push them down the ladder into hamburger-frying jobs. Cleverly these cultures are making their products scarce, by making their skills scarce. Push the frontiers of knowledge and there will always be higher prices because other nations lack the knowledge to compete.

Unfortunately Scott defined knowledge as science-based R&D. He measured complexity by the ratio of R&D to total product costs. But we could also argue that design equals the knowledge of arts applied to industry. The ever-better design is the infinite game which allows America to ascend the knowledge ladder. But let us now consider the second difference.

The Aim of the Finite Game Is to Win
The Aim of the Infinite Game Is to Keep Playing and Keep Improving

We have to somehow maintain the precarious balance between products and processes. Products are finite. They are obsolete or going out of fashion within weeks of production. While it is products or services which compete in finite games, it is processes—design processes, manufacturing processes, distribution processes, not to mention design for better manufacturing and distribution—which go on for ever. The infinite game is the process of improvement. The most important objective is for citizens of a society to go on playing, so they can improve and learn. The crime is to exclude them from this process to fester in ghettoes and pockets of despair.

Have you noticed that in Japanese, several words end in *do*: *judo*, *aikido*, *bushido*, *kyodo*? *Do* or *o* means "way of." The way of tea, the way of swords, the gentle way, the way of news, or simply as in Tao, "the way" or Shinto, "the way of gods." What is clear, I submit, is that in large parts of Asia the *process* is given priority over the *product*; the infinite game is considered more important than the finite game. Designing is given priority over the object designed. In the West the artist has to be safely dead before the price of his or her paintings soar. Scarce objects are valued above living skills.

A few years ago when John Sculley was head of Apple, he made a bold plea for process over product. He wrote in *Odyssey* that "the journey was the reward," that Apple was as old as Eden and as young as the human mind, that like Odysseus the company was journeying endlessly around the known world, dropping off its products at ports of call, to keep the odyssey going. What really mattered was the Knowledge

Navigator, a phantom ship treading the dawn of new knowledge. Few of our CEOs are so lyrical and imaginative. They prefer the adage: Winning is not the most important thing, but the only thing.

The problem with winning is simply stated, *the game comes to an end.* Suppose Rupert Murdoch, who recently dropped the price of the *Times* of London to 20p "won," and the *Independent* or the *Guardian*, two quality British papers, folded as a result. Would there be *more* competition or less? So while Murdoch is certainly "competing," he is doing so in a way designed to reduce diversity and competition in the longer run. He does not want the present number of quality papers "to go on playing." He does not want too many voices to contend. He wants his own to dominate.

This, then, is the contradiction at the heart of the finite game. It brings to a swift conclusion what it claims to value, namely competing. When the hostile takeover is launched, the merger concluded, the corporation acquired, then there are fewer players in the game. The oligopoly or duopoly is closer and market power has increased. Out of the wings and onto the stage stalks the Regulator, to the usual boos and hisses. We are all supposed to hate the government, but with competitors trying to eliminate their opposition permanently, governments have to step in and break up concentrations of power.

A close look at the Chinese, Japanese, or "tiger" economies shows a much larger number of competing companies. China has over one hundred pharmaceutical companies. Is this simply a prelude to the major shakedowns we witness in the West? Probably not, since Japan has eight automobile companies, for a population of 120 million, twenty fork-lift truck companies, forty machine-tool makers, and so on. Nor do most economies in Southeast Asia tolerate hostile takeovers. Why?

Because the ideal is to *go on playing* with as much variety and novel input as possible. As Akio Morito of Sony pointed out: *You do not break another man's rice bowl.* When your opponent is on the ropes, you stand back to let him recover. You do this because the ongoing *process* of competing is more important than the interim results. You do not let the results harm or impoverish the process. The company who lost round three may win round six as did Yamaha after being humbled by Honda. What is important is for an economy to keep coming up with the best ideas and the best designs, and for that you need multiple players, who learn from their mistakes without trauma or dislocation.

Good Guys versus Bad Guys or the Error-Correcting System

Players of finite games are haunted by the dualism of Good versus Evil. If I beat an opponent, then it follows that he was bad or worse than me, otherwise I was wrong to beat him. Classical economics teaches us that destroying opponents is good, because the resources under their economic management pass to the ownership of better managers. The allusion is to a Demolition Derby in which winners cannibalize the pieces falling off losing vehicles. There might be something in this if human beings were souped-up jalopies which did not bleed, or suffer, or despair; but smashing up a company has consequences that are not so benign.

But an interesting transformation occurs when we switch to the infinite game of continuous improvement. "Evil" turns into "error." "Good" turns into "correction," and we have an *error-correcting system* in which angels and devils are simply improvements and things needing further improvement. A corporation that does not classify at least 40 percent of what it does as "error" will not improve fast enough. Over two hundred years ago William Blake called for the marriage of heaven and hell. They meet and they marry in the game of *infinite improvement*.

I don't want to be soft or namby-pamby in this exposition. It is a dangerous world out there. Some finite games which lead to the defeat of attackers are inevitable. Our head-butting stags were being selected to hold their own against wolf packs. The point is not that we can end war or win-lose conflicts, but that the infinite game *includes many finite games within it*. If you improve and improve you can fight off external attacks. If you simply aim to win, you may cripple improvement in yourself and in your opponents. The finite games too often *exclude* the infinite.

Play Occuring within Boundaries with Definite Beginnings and Endings versus Play Occuring *within* Boundaries which Constantly Shift

In the finite game we know who the good guys are and where the boundary occurs between "us" and "them." Have you noticed how in economics most of the values that make life worth living, like compassion and a better environment, are referred to as "externalities"—things external to economics, of which design is just one more example. In the conventional view, the shareholders come first, and the corporation is a mere instrument of the purposes of its owners. Everyone else—customers, suppliers, subcontractors, employees, banks, partners, and the community—will be sacrificed if necessary, and they stand ready to sacrifice us to their interests.

We all know what may have to be done to meet Wall Street's profit forecast for the next quarter. A tenth of the employees may have to go, suppliers will have to take a 5 percent cut in payments or be replaced, consumers will have to pay the same for less, banks accept a debt moratorium or face default. It's one long struggle with all other units.

In the infinite game, you play *with* boundaries. For example banks, customers, and subcontractors are encouraged to hold shares in the company. You do not run your suppliers against each other, threatening to fire those with higher costs or more defects. You create a "shared destiny" relationship with single source suppliers, agree on the profits they should make, and then work together to make yours the best supply chain in the world, with the highest quality and the lowest costs.

These ever-more inclusive boundaries form what Michael Porter calls *clusters.* Individuals do not grow rich by themselves. Less and less do companies grow by themselves. What develops are *clusters* including a company, sub-contractors, suppliers, partners, banks, customers, and information infrastructures. You "play with the boundaries" to include valuable assets within these, just as Victor bested Sony by including more Hollywood films in its rental videos. The Japanese *keiretsu* is the cluster formalized.

In this view, even your competition is part of the cluster, continuously pushing up standards. High-wage demands by unions are part of the cluster, pressuring management to upgrade skills and invest more in each worker. When we play with boundaries to include more and more people, we all grow prosperous together. The leather-dyeing and shaping industry in Tuscany, Italy, is cited by Porter as a prospering cluster of small companies, led by leading fashion designers and firms. Silicon Valley is clearly a cluster as is Silicon Glen in Scotland. The infinite game pulls into its boundaries any new capabilities required. It is a shifting mosaic of value-added projects.

Those "playing with boundaries" may define fellow nationals as being insiders subject to friendly competition, yet define foreign nationals a outsiders. One suspects that the Japanese play an infinite game among themselves, yet finite games with us!

Competition as a Finite End in Itself versus Competing to Cooperate

In finite games you cannot have too much competitiveness since losing is final and winning is essential to survival. Victory requires the other's defeat. So important is winning that great hostility may be felt towards rules and umpires that make winning harder. The federal government shares the dismal reputation of referees at a football game, infuriating partisan spectators with their rulings. If you can cheat without being caught and penalized you do so. As the rules tighten to prevent this you simply escalate your abuse of the referee. The "Federal Register" has over four-thousand pages. In the end, the land of the free has the most regulations. And, of course, the number of lawyers increases exponentially.

In the infinite game, both competing as a process and cooperating as a process are phases of an overall need to learn. When we compete we differentiate our offerings. When we cooperate we integrate and share the best solutions which the competitive phase threw up. We learn by progressively differentiating and integrating, by competing and then cooperating. These need to be fine-tuned. We're back to the Dionysian harmony and *symphoniasis* of classical Greece.

Let me illustrate how this works in Motorola. Throughout this transnational company, TCS groups form spontaneously and self-organize. TCS stands for Total Consumer Satisfaction. These groups provide consultations to their own departments by diagnosing faults and pioneering improvements. They do not simply advise like a consultant. They demonstrate their propositions. They compete worldwide in around a dozen categories, judged by the value as well as the creativity of their initiatives.

Heats take place within regions of a nation, within nations, within regions of the world. Finally three to four hundred groups stage their finals in the Paul Galvin Center in Shaumburg, Illinois. There is a carnival atmosphere. Many competing groups wear national costumes and call themselves The Oriental Express or the Boys from Bangalore. Every solution celebrated is shared with all contestants. All savings achieved or quality records are flashed on an electronic board. Different cultures are encouraged to introduce themselves with a video.

The competition is entirely friendly. The information then shared is wholly cooperative. Instead of each employee competing in an American game, each culture shows what it can do using its own values. This is a game of games, an infinite game, where cooperating and competing are woven finely and the rules themselves are being negotiated. I come now to the last major difference.

Winners of Finite Games Live On in the Memory of Witnesses Contributors to the Infinite Game Live On in the Patterns of Play Itself

In the finite game we dream of joining the Hall of Fame. The victory must be visible, clear-cut, dramatic, and physical. For the likes of O. J. Simpson, nothing will be that simple again. Sheer physical prowess will not be enough. Our yearning for victories in finite games is too crude, too stereotyped, too macho. It has the culture of the locker room. Long ago we came up against the limits of push and shove. The trick is not to work harder, but to work smarter. You work smart when you learn.

We cannot all be heroes on some imagined football field. It requires a ratio of stars to audience of one to a thousand. We must outgrow such childlike dreams which doom most of us to disappointment. Fame is for very few of us, but nearly all of us can live on in the patterns of play itself. Participants find an immortality of a kind in weaving together the threads of an infinite design. For finally we do not simply design products, or logos, or buildings, we design values and social processes. We self-organize to solve problems, and in what we create, the values we have reconciled live on, erring and correcting, competing and cooperating, differentiating and integrating, process and product, community and privacy, changing and continuing.

Many years ago a female anthropologist and poet, Ruth Benedict, did a study of three American Indian tribes. Two were miserable, self-destructive, and depressed. One was relatively healthy and vibrant. She studied forty different variables and discovered to her horror that a lifetime of social science could not even distinguish heaven from hell. She was very nearly at the end of her life and never completed her report.

But in her last months, she did radically revise her concepts and escaped from the finite game of Newtonian science. We had been asking the wrong questions, she wrote. The difference between the tribes was not whether they were predominantly self-interested or selfless in their values. The difference was that in one tribe egoism and altruism were *synergistic*, were socially designed to work together, whereas in the other tribes these values were polarized. Selfless behaviors were exploited, so all self-interest was sanctioned in fierce finite contests where those who succeeded were socially punished. Perhaps it was the poet in her, perhaps the woman, perhaps the anthropologist, but Ruth Benedict came up with the idea of *synergy*, as a form of social design that transcended the elements of which it was comprised, a transformational synthesis, a leap from the finite to the infinite game.

Her working paper was discovered by Abraham Maslow, the psychologist. I still remember the paragraph in his 1954 book *Motivation and Personality* in which he described the psychological mind-sets of great Americans, *self-actualizing persons*, he called them. That passage glowed in the dark for me. It still does. He wrote:

> The age-old opposition between heart and head . . . was seen to disappear where they became synergic rather than antagonistic . . . the dichotomy between selfishness and unselfishness disappears. . . . Our subjects are simultaneously very spiritual, and very pagan and sensual. Duty cannot be contrasted with pleasure nor work with play when duty *is* pleasure. . . . Similar findings have been reached for kindness-ruthlessness, concreteness-abstractness, acceptance-rebellion, self-society, adjustment-maladjustment . . . serious-humorous, Dionysian-Apollonian, introverted-extroverted, intense-causal . . . mystic-realistic, active-passive, masculine-feminine, lust-love, and Eros-Agape . . . [all] coalesce into an organismic unity and into a non-Aristotelian interpenetration . . . and a thousand serious philosophical dilemmas are discovered to have more than two horns, or paradoxically, no horns at all.

Synergy, as you all know, was taken up by Buckminster Fuller and made a major principle, not simply of architecture, but of reordering social systems. It was what Dick Farson called "meta-design," the game of games, design of designs.

This is becoming a familiar theme. Americans invent ideas, like W. Edwards Deming's notion of the error-correcting system, Joseph Scanlon's vision of the self-organizing team of workers in the Scanlon Plan (an idea extolled by Douglas McGregor as early as 1959), but foreigners pick these up and incorporate them before we do. There are thousands of Scanlon Plans in Japan. Why? Because the infinite game is encoded in Eastern philosophy, in the Tao, in the t'ai chi, in Zen koans and riddles, in the Mandala, and in the cycle of eternal return.

We tend to classify these as mystical and non-rational, but in truth they have the encompassing reason of open systems and creative dialogue. We enjoy brief glimpses of transformational leadership—the campaigns of Martin Luther King—before sinking once again beneath the sheer weight of scientism and technical reasoning, reconstructed *after* the event and displacing the logics of creativity and actual discovery. We create but have no logic of creation and so are forced over and over again to start from the beginning, while our competitors pass the baton in that great relay race which is the infinite game.

We speak in electronics of first-, second-, third-, and fourth-generation products, but then we shrink from the implications of this metaphor, which is that electronics, like stags, like human beings, pass on genetic information from generation to generation. It is Lovejoy's Great Chain of Being which Newtonian science destroyed. The Great Chain of Being is the infinite game, wherein whole nations invest in ever-escalating skills and in the logics that join generations.

It is a system in which, as Stewart Brand put it, "Where generosity leads, prosperity follows." It's like parents who give love to their children, not so that they'll give it back, not because they're getting interest on it, but because of the next generation of people, the next generation of products. As Larry Keeley would say, "You transform and you reinvent." You transcend.

Surely design professionals, such as you, are in the best position to deliver the message, *transcend or die*. Break off the sterile jousts of finite games, the world of liar's poker, the Michigan Militia, the paranoid hatred of cultural diversity that has us in its grip, and join the dance-that-invents-itself, the infinite game.

Concluding Comments
John Kao

It is now appropriate to raise the question: *Quo vadis?* Where do we go from here? Milton Glaser has said, "The war is over and business has won." I have modified my position as a result of the discussions we've had and now conclude that both sides are positioned to win in what I now know to be the infinite game. But there is much to be done on the relationship between design and business to make sure that it enjoys the benefits of creative abrasion and not merely abrasion.

Change is inevitable in considering the relationship and integration of design and business. It is a challenge to think about design as a core competency in companies and not merely as a short-order meal. It is a challenge to think about how to be a designer at a time of rapid change, what the Global Business Network calls a "scenario of perpetual transition." Also, it is important to recall that the question is not about design, *per se*. In business, design is presold; it is an expanding arena of concern. The issue is designers and their positioning. The traditional models of organization may not be adequate. There are many crucial questions in an environment where change is a fact not a choice. Design cannot stay still. And if designers are creators and keepers of the shared space, as Michael Schrage has said, what happens when that expertise becomes more of a commodity, when businesses are learning to perform many of these functions for themselves?

What are the new strategic imperatives for designers? What is the functionality of being big versus small or a full-service company versus a boutique? Should designers become strategy consulting firms, Innovations "R" Us? How much design will be done by companies themselves? How much by professional design firms? How much of design will be incorporated into organizational processes?

How much should designers know about business? Is it true, as *Business Week* has suggested, that the changing corporate needs *are* forcing flamboyant designers to become serious business people? How does the designer deal with the role of being an entrepreneur? How should a designer optimize his or her position? How should a designer think about issues of forward and vertical integration to capture value?

What form of organization is the optimal design firm of the future? Is competing through scale still a viable strategy? Should the domain of vertical integration be expanded to include market research and other functions? What can be done through the technology of outsourcing? This is especially important in an environment where the designer's awareness must always exceed that of their client. This involves integrating new technology with the imperatives of collaboration.

What should new relationships with talent be? What are the new clustering of relationships that revolve around the twin axis of permanent employment versus temporary alliances? What are new ways of gathering talent?

What are new ways of adding value in the design process? I maintain that design is a Trojan horse for the overall agenda item of creativity in business. Designers can help organizations to respond and to create, not merely to react. They can help maintain an awareness of their environment and of the importance of creativity.

Where do we go from here? In charting a path to genuine creativity, it is useful to recall that the Zen Buddhists call for the purity of a "beginner's mind." Tom Peters echoed this in his speech when he said, "What is design? I don't know." This reminds us that the definition of design in the nineties is up for grabs. Part of design mindfulness is keeping such questions in mind. If the success of this conference is measured by the quality of the questions that it has aroused and the quality of the conversations that it has stimulated, then I will feel satisfied by its outcome.

I'd like to express my appreciation to the speakers for their confidence in stepping into the unknown. This conference took up the question of defining the new business of design and, I think, took a big bite out of the topic. It was a great jam session and the music sounded fine. Thank you from the bottom of my heart for being here.

The Participants

Jane Alexander

Jane Alexander was nominated in 1993 by President Bill Clinton to become the sixth Chairman of the National Endowment for the Arts. Unanimously confirmed by the U.S. Senate, Alexander has since visited more than 130 communities in all 50 states and Puerto Rico to address topics including the contributions of art to educational reform and community building. Under her leadership, the Arts Endowment has initiated new partnerships with the U.S. Department of Transportation to encourage design excellence at the federal level, the Corporation for National Service to create Writers' Corps, part of the national force of AmeriCorps volunteers, and with Canada and Mexico to begin a three-way artistic exchange program. In 1994, Alexander convened "Art 21: Art Reaches into the 21st Century," the first federally organized national arts conference, which drew together more than 1,100 artists and administrators to explore the arts' and the federal government's conjoining role in the next century.

For her leadership, Jane Alexander has received honors including the 1988 Living Legacy: Jehan Sadat Peace Award, the 1995 Montblanc de Culture North America Award, and most recently a Common Wealth Award. During her thirty-five-year career as an actress, producer, and author, she appeared in more than forty films and one hundred plays, earning a Tony for "The Great White Hope," an Emmy for *Playing for Time*, and the Television Critics Circle award for her portrayal of Eleanor Roosevelt in *Eleanor and Franklin: the White House Years*. A Boston native, Alexander is the granddaughter of Daniel Quigley, who was Buffalo Bill's personal physician in North Platte, Nebraska.

Stewart Brand

In 1966, ex-Army officer Stewart Brand designed buttons on which he posed the question: "Why Haven't We Seen a Photograph of the Whole

Earth Yet?" His campaign provided the impetus for NASA to make good color photos of the earth some two years later. The ecology movement took shape partially as a result of these photos.

Brand went on to found and edit the original *Whole Earth Catalogue* and *The Last Whole Earth Catalogue* which received the National Book Award in 1972. In 1974, he wrote *Two Cybernetic Frontiers*, a book about cutting-edge computer science which contained the first use of the term "personal computer." Ten years later, Brand founded the WELL (Whole Earth 'Lectronic Link), a computer teleconference system for the San Francisco Bay area which now has ten thousand active members worldwide. He also initiated and organized "The Hacker's Conference," which has been held annually since 1986.

In 1988, he co-founded the Global Business Network, a consulting organization that explores global futures and business strategy for sixty multinational corporations including Bell South, IBM, Xerox, Unocal, ABC, and Nissan.

He is the author of *The Media Lab: Inventing the Future at MIT* and *How Buildings Learn: What Happens After They're Built*. He is also the recipient of the Golden Gadfly Lifetime Achievement Award from Media Alliance, San Francisco.

David Carson

Sociologist, surfer-turned-graphic designer, David Carson is art director of *Ray Gun* magazine, which called itself "The Bible of Music and Style" until a critic equated Carson's typography with "the end of print." Now *Ray Gun* calls itself "The Bible of Music and Style and the End of Print." Although he has not yet ended print, Carson is highly responsible for beginning some arguments about it. His highly acclaimed and highly controversial work has been featured internationally in such publications as *Creative Review* (which called him "art director of the era"), *Emigre* (which devoted an entire issue to him), *Graphis*, *I.D.*, and *USA Today*. The International Center for Photography named him "Designer of the Year." David Carson's clients include Nike, MTV,

Lotus, David Byrne, the artist formerly known as Prince, American Express, Pepsi, Ryder Trucks, Glendale Federal Bank, and Levis.

Bran Ferren

Bran Ferren is a designer who works in theater, film, music, product design, architecture, and the sciences. He is Executive Vice President for Creative Technology of Walt Disney Imagineering (WDI), the theme park master planning, research and development, design engineering, production and project management subsidiary of the Walt Disney Company. His responsibilities include overseeing all of Imagineering's research and development activities on both coasts. His R & D mission also extends beyond Imagineering to include the development of technologies that could have strategic value throughout the Walt Disney Company.

Prior to Disney, Ferren headed his own firm, Associates & Ferren, which specialized in research and development, creative design, engineering and execution of projects for the visual and performing arts, as well as for industry and the sciences. Ferren produced and directed the feature film, *Funny*, and directed the special visual effects for *Altered States, Deathtrap, Places in the Heart, Making Mr. Right, The Manhattan Project, Star Trek V: The Final Frontier,* and *The Little Shop of Horrors,* for which he was a 1987 Academy Award nominee. A theatrical designer for such Broadway shows as *Evita, Cats,* and *Sunday in the Park with George,* he has also worked with rock musicians Pink Floyd, Depeche Mode, David Bowie, and Paul McCartney.

Milton Glaser

A long-time board member and past president of IDCA (1990–91), Milton Glaser is widely known for his work in the design of print graphics, subway stations, restaurants, museums, and landmarks like the observation deck and permanent exhibition for the Twin Towers of the World Trade Center. An expert in redesign, he has redesigned magazines, supermarkets, and the reputation of New York. Over the course of forty years, Glaser helped found Push Pin Studios; *New York* magazine; WBMG, a publication design firm; and his present design studio, Milton Glaser Inc. The logo for Tony Kushner's Pulitzer Prize winning play, *Angels in America*, is the basis of one of over three hundred posters for which Glaser is responsible. His work has been exhibited in Europe and North America and is in the permanent collections of the Museum of Modern Art, The Israel Museum, The Chase Manhattan Bank, and the National Archive, Smithsonian Institute. He is a board member and instructor at the School of Visual Arts, New York. Glaser is the recipient of the following awards: The Society of Illustrator's Gold Medal, The St. Gauden's Medal from Cooper Union, Metro International's Fulbright Award for Individual Achievement, and the American Institute of Architect's Institute Honor.

David Gresham

David Gresham is a Vice President in Product Development at Fitch Inc. While there, he has worked on product design and systems development for Eastman Kodak, Hitachi, Xerox, AT&T, Digital, Mitsubishi, NCR, Thermos, and Steelcase. Prior to joining Fitch, Gresham was Director of Design for Details, a Steelcase Design Partnership venture where he was responsible for all aspects of the company's image as well as all product development.

Before Details, Gresham was principal and co-founder of Design Logic, a product development/advanced research firm. He was also the corporate manager of industrial design for ITT Corporation and senior associate designer at IBM.

Charles Hampden-Turner

Charles Hampden-Turner is a Permanent Visitor at the Cambridge University Judge Institute of Management Studies and a professor on extended leave from the Wright Institute in Berkeley, California. He has written eleven books, including: *Seven Cultures of Capitalism, Corporate Culture: Vicious and Virtuous Circles*, and *Charting the Corporate Mind*.

His best-selling *Maps of the Mind* was a Book-of-the-Month-Club selection and has sold over 150,000 copies worldwide.

As a consultant, he has worked with Apple, Barclay's Bank, British Airways, British Petroleum, Clorox, Coopers & Lybrand, IBM, Nissan, Shell, and others. He worked for Group Planning at Shell and then became Shell Senior Research Fellow at the London Business School.

Hampden-Turner earned his masters and doctorate from Harvard Business School. He is a winner of the Douglas McGregor Memorial Award and the Columbia University Prize for the Study of the Corporation.

Craig Hodgetts

Architect Craig Hodgetts is creative director for Hodgetts + Fung Design Associates in Los Angeles. His work as an architect and urban planner includes an award-winning master plan for Arts Park L.A., a project that featured a comprehensive site plan as well as the architectural design for a Natural History Museum and Media Center; new construction and renovation of the Craft and Folk Art Museum of Los Angeles; and the award-winning UCLA "Towell" Library and master plan.

Hodgetts has also designed entertainment venues such as theme parks for Sony, MCA, Universal, and Tokyo Disneyland. He is involved with the conceptual design of Deep Rock Drive, a chain of rock-and-roll micro theme parks, and with the installation design of *Art and Film*, an exhibition scheduled to open this fall at the Museum of Contemporary Art, Los Angeles.

Hodgetts is a three-time recipient of *Progressive Architecture*'s First Design Award and a recent recipient of the Architecture Award from the American Academy of Arts and Letters. He is a professor at UCLA's School of Architecture and Planning.

John Kao

John Kao is a designer of enterprises. He is an author, lecturer, and practitioner in the intersecting fields of corporate creativity, information technology, and entrepreneurial management. His training is in both psychiatry (Yale and Harvard medical schools) and business (Harvard Business School).

He is author of the book series *Managing Creativity, The Entrepreneur,* and *The Entrepreneurial Organization* published by Prentice Hall. His latest book, *Jamming: The Art and Discipline of Corporate Creativity*, is published by HarperCollins. His research on Chinese entrepreneurs was published in the *Harvard Business Review* as "The Worldwide Web of Chinese Business."

For the past twelve years, Dr. Kao has taught creativity courses at Harvard Business School in the M.B.A. and Advanced Management programs. He is Program Chair for the school's executive program, Enhancing Corporate Creativity. He has also taught at Yale, The University of Copenhagen, and the Stanford Graduate School of Business.

Dr. Kao is CEO of The Idea Factory, a company developing audiovisual software for business and technology-based tools that enhance corporate creativity. He is also founder of several other companies including: Genzyme Tissue Repair, a public company involved with tissue engineering; K.O. Technology, next generation cancer diagnostics and therapeutics; and Pacific Artists, a feature film production company. Dr. Kao is an advisor to companies in the fields of biotechnology, computer software, entertainment, and financial services. He has also consulted to the World Health Organization and to UNESCO.

Dr. Kao was production executive on *sex, lies and videotape*, which won the Palme d'Or at the 1989 Cannes Film Festival. He was also executive producer of *Mr. Baseball*, a Fred Schepisi film starring Tom Selleck that was financed and distributed by MCA/Universal. Dr. Kao had a "two-second" cameo in Robert Altman's *The Player* and is currently working on several feature film projects himself. He is a member of the

Global Business Network and a 1995 Fellow of the World Economic Forum. In his spare time, he plays jazz piano.

Mitchell Kapor

Mitchell Kapor is Adjunct Professor in Media Arts and Sciences at MIT and is affiliated with MIT's Media Lab. In 1990, he co-founded the Electronic Frontier Foundation (EFF). Based in Washington, D.C., the EFF is a nonprofit, public interest organization that studies the social impact of digital communications media in a free society. He founded the Lotus Development Corporation and designed the Lotus 1-2-3 software program. Kapor served as president, CEO, and later Chairman of Lotus. He later served as Chairman and CEO of ON Technology, a developer of software applications for workgroup computing. In 1992-1993, he chaired the Massachusetts Commission on Computer Technology and Law which was chartered to investigate and report on issues raised by computer crime in the state. He has also been a member of the Computer Science and Technology Board of the National Research Council. Kapor currently serves on the National Information Infrastructure Advisory Council.

Larry Keeley

Larry Keeley is president of Doblin Group, the Chicago strategic design planning firm, where he has spent nearly sixteen years focusing on the strategic value of design for large corporations.

Keeley pioneered the field of strategic design planning, a unique combination of conventional strategic planning, social science research, and user-centered design. This practice involves simulation of new business opportunities, offering powerful illustrations of new strategies with examples of future products, services, communications, and information systems.

Keeley is active in creating special methods for integrated design programs, intended to help clients achieve massive industry breakthroughs. Doblin Group clients have included: Aetna, American Hospital Supply, Amoco, Beatrice Foods, Hallmark, McDonald's, Steelcase, Texas Instruments, Xerox, and others.

A Board Member of the American Center for Design and a past Board Member of the Design Management Institute, Keeley is also on the Board of Overseers for the Institute of Design, Illinois Institute of Technology, where he serves as an associate professor teaching graduate courses in design strategy.

He is a member of the Governing Board of Directors for WBEZ, public radio in Chicago.

Dorothy Leonard-Barton

Dorothy Leonard-Barton, the William J. Abernathy Professor of Business Administration, joined the Harvard faculty in 1983 after teaching for three years at the Sloan School of Management, Massachusetts Institute of Technology. She has taught MBA courses in manufacturing, new product and process development, technology strategy, and technology implementation. At Harvard, MIT, and for corporations such as Digital Equipment, AT&T, and Johnson & Johnson, she has conducted executive courses on a wide range of technology-related topics such as cross-functional interfaces and technology transfer during new product and process development. She served for three years as faculty chair of the Harvard course on Building Development Capabilities, a new program for upper-level executives on managing new product and process development. She is also a faculty member in the Harvard programs Managing Global Opportunities and Managing International Collaboration.

Professor Leonard-Barton's major research interest and consulting expertise are in technology strategy and commercialization; she is currently studying the design and use of multimedia communication linking members of new product development teams. She has consulted with and taught about technology management for the Swedish and other governments and for large corporations such as Kodak. She serves on the corporate Board of Directors for American Management Systems, an industry leader in custom software development. Her numerous articles have appeared in such journals as *Design Management Journal, Strategic Management Journal, Harvard Business Review,* and in books on technology management. Her book, *Wellsprings of Knowledge: Building and Sustaining Core Technological Capabilities* will be published in 1996 by Harvard Business School Publishing.

Tom Peters

Described by *Business Week* magazine as business's "best friend and worst nightmare," Tom Peters is an author whose books have been found at or near the top of *The New York Times* bestsellers list since 1982's *In Search of Excellence.* Among his other titles are *A Passion for Excellence, Thriving on Chaos* and *Liberation Management.* In response to current business conditions, he is writing a series of paperback originals beginning with 1994's *The Tom Peters Seminar: Crazy Times Call for Crazy Organizations* and *The Pursuit of Wow!: Every Person's Guide to Topsy-Turvy Times.* Peters writes a bi-monthly column for *Forbes ASAP.* He has also written over a hundred articles for various publications including: *Business Week, The Economist, The Financial Times, The Wall Street Journal, The New York Times,* and *The Harvard Business Review.*

As a speaker, Peters hosts fifty seminars a year in places as far flung as Australia, England, Malaysia, India, Brazil, and Dubai. He has created and starred in numerous corporate training films as well as hosted award-winning documentaries for United States Public Television. He is founder of The Tom Peters Group, three training and communication companies in Palo Alto, California.

Samina Quraeshi

Artist, designer, and author Samina Quraeshi is the Director of the Design Arts Program of the National Endowment of the Arts. Raised in Pakistan and educated in the United States, Quraeshi was Assistant Director of the Carpenter Center for the Visual Arts at Harvard University and principal in the design firm Shepard Quraeshi Associates in Chestnut Hill, Massachusetts.

Quraeshi has worked in design studios in London, New York, Denver, and Boston and has taught at Yale, Boston University, and the Rhode Island School of Design. She has served as a design consultant to the government of Pakistan, founded *Focus on Pakistan*, a quarterly magazine featuring the people and culture of Pakistan, and established a workshop for women in Pakistan dedicated to reviving traditional textile arts. She has had solo exhibitions of her art at Harvard, MIT, The Smithsonian, Zamana Gallery in London, and the Kansas City Art Institute. Quraeshi has received numerous awards, among them the Gold Medal "Best of the West" for her work on the Bicentennial of the State of Colorado, the Distinguished Designer Award, and the Medal for the Arts from the government of Pakistan.

Michael Schrage

Michael Schrage writes and consults on the ways technology reshapes relationships in organizations. He explores collaborative design and new media technologies as a research associate with the MIT Sloan School's Center for Coordination Science and the MIT Media Lab. He is the author of *Shared Minds: The New Technologies of Collaboration* and currently writes the syndicated weekly "Innovation" column for the *Los Angeles Times*; *AdWeek*'s fortnightly "Out There" column; *Computer-World*'s monthly "Counter-Information" column; and has contributed to the *Harvard Business Review, The Wall Street Journal, Wired, I.D.,* and the *Design Management Journal*. His article, "The Culture(s) of Proto-typing," won the Doblin Prize for best article in the *Design Manage-ment Journal*. He was formerly a senior editor at *Manhattan, Inc.* magazine and the technology correspondent for the *Washington Post.*

Schrage's research focuses on the design and development of proto-types, collaborative tools, and organizational media to support innova-tion. He consults to several firms in these areas including: Allstate, NTT, Coopers & Lybrand, IDEO, PacTel, Steelcase, and the Goldhirsh Group, where he created the FaxPoll for *Business Month* magazine.

235

Jivan Tabibian

Jivan Tabibian is a social scientist and urban planner. As an educator he has taught political science, urban design and planning, and management at various universities including UCLA and USC. As a consultant, he has experience in management, marketing, and public- policy planning, as well as land-use planning. He has served on the board of Directors for the Sundance Institute for Film and Television, IDCA, and several private corporations. He lectures internationally on design theory, community development, and cultural criticism.

His passion for food, drink, and cigars has saddled him with a restaurant, REMI, in Santa Monica, California.

John Thackara

In 1993 John Thackara was appointed the first Director of the Netherlands Design Institute in Amsterdam, where he recently chaired the first European Design Summit and helped inaugurate the Doors of Perception conferences on the design challenge of interactivity. Prior to that, Thackara was a management consultant and design producer who helped business, higher education, and the arts work profitably together. His clients have included companies (IBM, Japan Airlines, Toyosash, Comit Colbert, Rover, Dunhill); government agencies (The European Commission, Glasgow Develpment Agency, MITI of Japan); and academic institutions (The Royal College of Art, where he was director of research for three years, and the Architectural Association). Since 1985, he has lectured and run workshops on architecture, design management, and new media strategies in twenty-four countries. He has also curated exhibitions at the Centre Georges Pompidou in Paris, the National Museum of Art in Kyoto, and the AXIS gallery in Tokyo. Thackara has written for numerous national newspapers and is the author of the following books: *New British Design, Design After Modernism*, and *Lost in Space*.

Hatim Tyabji

The simple swipe of a credit card at a gas station begins a sequence of electronic events which, chances are, have been scripted by Hatim Tyabji. Tyabji is CEO of VeriFone, a corporation that specializes in worldwide transaction automation for retail merchants, petroleum service stations, convenience stores, supermarkets, healthcare providers, and government benefits agencies. VeriFone has shipped more than four million systems, supporting over 1400 applications to customers in over 90 countries. Tyabji is also the chief author of a set of written values called The VeriFone Philosophy. It is under these values—which help define VeriFone's corporate culture—that this "virtual" company operates with over 1900 employees in more than 30 offices around the world. Prior to VeriFone, Tyabji was president of the Information Technologies Group of the merged Sperry and Burroughs organizations, which have since become the Unisys Corporation.

Lorraine Wild

Lorraine Wild is a partner at ReVerb, a graphic design firm in Los Angeles. Her design work has included collaborations with architects, museum curators, and publishers in the United States and abroad. Her work has appeared in *I.D., Print, Design Quarterly, Studio Voice,* and in the recent book, *The Graphic Edge.* Her writing has appeared in *Emigre, I.D., Print, Graphic Design in America, Cranbrook Design: The New Discourse, Lift & Separate,* and *The Edge of the Millennium.* She is on the faculty of the design program at the California Institute of the Arts, which she directed from 1985 to 1991. Wild serves as a project tutor for the graduate program at the Jan van Eyck Akademie in Maastricht, The Netherlands.

Lorraine Wild has received design awards from The American Center for Design, the AIGA, and the American Association of University Publishers and was named as one of *I.D.*'s "Top Forty" in 1993.

Acknowledgements

The IDCA is deeply grateful to the following organizations and individuals for their generous support and unique contributions to the 1995 program:

Hallmark Cards, Inc. for their continuing sponsorship, placing them among our most loyal and generous benefactors;

Siemens AG for their support of the 1995 conference and their early and active commitment to the 1996 program;

Masami Shigeta for his exceptional efforts in helping us break new ground by arranging and underwriting the simulcast to Tokyo.

For their generous support of the student scholarship program, IDCA wishes to thank:

Henry Schein, Inc.
Schein Pharmaceutical Inc.

For their generous contributions:

Alessi
Alexander Julian Foundation
AT&T
BMW of North America, Inc.
Champion International Corporation
Gilbert Paper
Hallmark Cards, Inc.
Henry Schein, Inc.
Macromedia
Mead Coated Papers

Nissan
Philips Corporate Design
Polaroid Corporation
Schein Pharmaceutical Inc.
Siemens AG
Unifor Inc.
Whirlpool

Special thanks are due to the **Lotus Corporation** and to **Lante Corporation** for their contributions to the virtual conference that ran concurrent with the main conference; to **Hiroko Sakamura**, president of **Transform Corporation**, **Kiyoshi Awazu**, and **Ikko Tanaka** for their efforts in arranging the simulcast of the conference to Japan; to **PictureTel Corporation** for providing the video transmission for both the simulcast to Japan and the videoconference to New York during the roundtable discussion; and to the **Aspen Institute** for the special support that they have given the IDCA through the years.

Thanks also to Lex Lalli, Deborah Murphy, Karen Petersen, and the Aspen volunteers; to Wendy Keys for input on the film program; to Ralph Caplan, Tom Hardy, and Saul Bass for their input; and to Sally Andersen-Bruce for running the Cafe.

International Design Conference in Aspen

Board of Directors

Richard Farson, *President*
Paola Antonelli
Saul Bass
Julian Beinart
Marshall Brickman
Ralph Caplan
David Carson
Ivan Chermayeff
Lou Dorfsman

Bran Ferren
Nancye Green
Alexander Julian
Larry Keeley
Wendy Keys
Noel Mayo
William Stumpf
Harry Teague
Jane Thompson

Board of Advisors

Sally Andersen-Bruce
Kiyoshi Awazu
Robert Blaich
Gordon Bruce
Jay Chiat
Michael Crichton
Andrew Drews
Helene Fried
Milton Glaser
Tom Hardy

Rosamind Julius
Margaret Marshall
Louis Nelson
Dianne Pilgrim
Hiroko Sakomura
Adele Naudé Santos
Masami Shigeta
Frank Stanton
Jivan Tabibian
Ikko Tanaka

I Index

I